Science and Engineering of Freak Waves

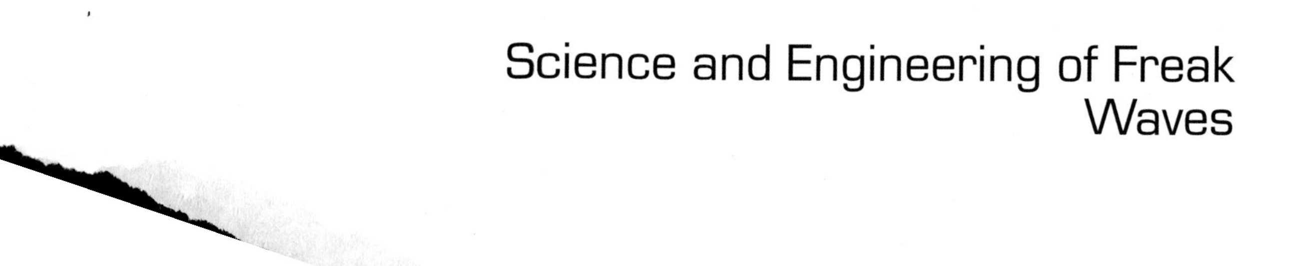

Science and Engineering of Freak Waves

Edited by

Nobuhito Mori
Disaster Prevention Research Institute, Kyoto University, Japan

Takuji Waseda
Graduate School of Frontier Sciences, The University of Tokyo, Japan

Amin Chabchoub
Disaster Prevention Research Institute, Kyoto University, Japan

ELSEVIER

Elsevier
Radarweg 29, PO Box 211, 1000 AE Amsterdam, Netherlands
The Boulevard, Langford Lane, Kidlington, Oxford OX5 1GB, United Kingdom
50 Hampshire Street, 5th Floor, Cambridge, MA 02139, United States

ISBN: 978-0-323-91736-0

For Information on all Elsevier publications
visit our website at https://www.elsevier.com/books-and-journals

Publisher: Candice G. Janco
Acquisitions Editor: Maria Elekidou
Editorial Project Manager: Sara Valentino
Production Project Manager: R. Vijay Bharath
Cover Designer: Vicky Pearson Esser

Typeset by MPS Limited, Chennai, India

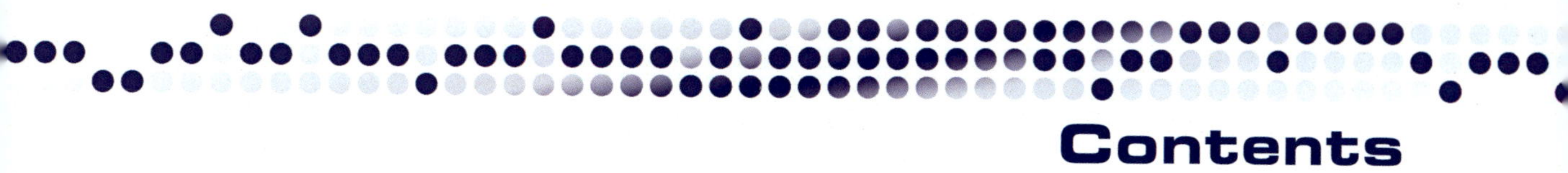

Contents

List of contributors

Francesco Barbariol Institute of Marine Sciences (ISMAR), National Research Council (CNR), Venice, Italy

Alvise Benetazzo Institute of Marine Sciences (ISMAR), National Research Council (CNR), Venice, Italy

Filippo Bergamasco Institute of Marine Sciences (ISMAR), National Research Council (CNR), Venice, Italy; University of Venice "Ca' Foscari,", Venice, Italy

Jean-Raymond Bidlot ECMWF, Reading, Berkshire, United Kingdom

Elzbieta M. Bitner-Gregersen DNV, Høvik, Norway

Amin Chabchoub Disaster Prevention Research Institute, Kyoto University, Japan; School of Civil Engineering, The University of Sydney, Sydney, NSW, Australia

Marios Christou Civil & Environmental Engineering Department, Imperial College London, London, United Kingdom

Dong-Jiing Doong Department of Hydraulic and Ocean Engineering, National Cheng Kung University, Tainan, Taiwan

Kevin Ewans MetOcean Research, New Plymouth, New Zealand; Department of Infrastructure Engineering, University of Melbourne, Melbourne, VIC, Australia

Hidetaka Houtani School of Engineering, The University of Tokyo, Bunkyo, Tokyo, Japan

Suzana Ilic Lancaster Environment Centre, Lancaster University, Lancaster, United Kingdom

Hiroaki Kashima Coastal and Ocean Development Group, Port and Airport Research Institute, Yokosuka, Kanagawa, Japan

Jamie Luxmoore Lancaster Environment Centre, Lancaster University, Lancaster, United Kingdom; Orcina Ltd., Ulverston, United Kingdom

Zuorui Lyu Disaster Prevention Research Institute, Kyoto University, Kyoto, Japan

Nobuhito Mori Disaster Prevention Research Institute, Kyoto University, Japan

List of contributors

Chuen-Teyr Terng Marine Meteorology Center, Central Weather Bureau, Taipei, Taiwan

Cheng-Han Tsai Department of Marine Environmental Informatics, National Taiwan Ocean University, Keelung, Taiwan

Takuji Waseda Graduate School of Frontier Sciences, The University of Tokyo, Japan

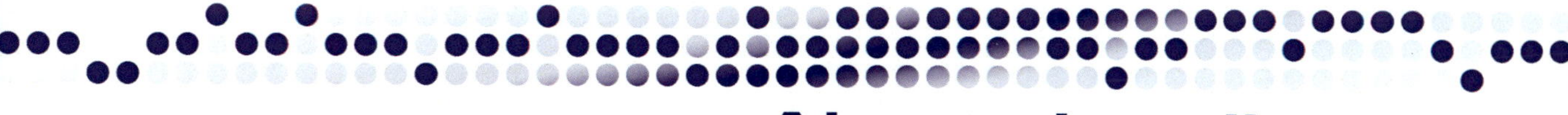

About the editors

Nobuhito Mori

Disaster Prevention Research Institute, Kyoto University, Japan

Nobuhito Mori is currently the deputy director of the Disaster Prevention Research Institute at Kyoto University, Japan, and also an honorary professor at the Swansea University, United Kingdom. He received his PhD from the Gifu University, Japan in 1996 and since then has had several posts at leading institutes. Professor Mori's research interests include air–sea interface physics, dynamics of wind waves, long waves and tsunamis, tropical cyclones and related coastal disasters, and nonlinear wave dynamics. He has won numerous awards in the last 20 years but, most recently, he has been awarded the 2012 Prize of Minister of Science and Technology in the Commendation for Science and Technology by the Minister of Education, Culture, Sports, Science and Technology of Japan and the 2021 Gambo-Tatehira Award by the Meteorological Society of Japan. Professor Mori has published numerous articles, book chapters, and conference proceedings.

Takuji Waseda

Graduate School of Frontier Sciences, The University of Tokyo, Japan

Takuji Waseda is currently a professor in the Department of Ocean Technology Policy and Environment at the University of Tokyo. He received his PhD in 1997 at the University of California, Santa Barbara. Waseda's research includes observational, experimental, and numerical studies of freak waves for the safety and efficiency of ship navigation. He has also conducted research on waves, ocean currents, and thermal energy resource assessments. Recently, he became active in research on wave–ice interaction for the support of vessels navigating in the Arctic and Antarctic Oceans. He was a member of the 64th Japanese Antarctic Research Expedition. Professor Waseda has published numerous articles, book chapters, and conference proceedings.

Amin Chabchoub

Disaster Prevention Research Institute, Kyoto University, Japan

Amin Chabchoub received his PhD from the Hamburg University of Technology, Germany, in 2013 and is currently an associate professor at the Kyoto University's Disaster Prevention Research Institute and at the University of

Sydney's School of Civil Engineering. His areas of expertise and research interests include wave hydrodynamics, environmental fluid mechanics, nonlinear dynamics, natural hazards, and extreme events. Associate professor Amin Chabchoub has published a variety of interdisciplinary articles, book chapters, and conference proceedings.

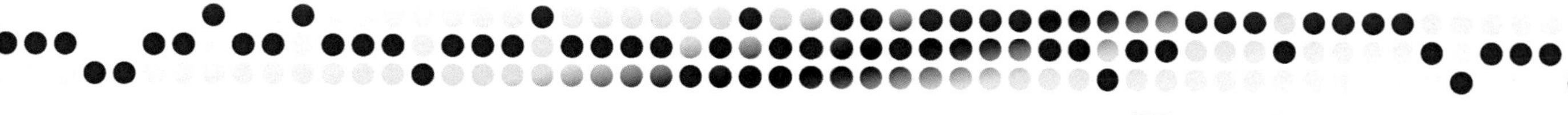

Preface

Extreme waves in the ocean, also referred to as freak or rogue waves, have been extensively studied since the early 2000s. Giant extreme waves have severely damaged and, in some cases, even sunk ships, demolished offshore oil field facilities and offshore wind turbines, and were also responsible for several fatalities. The existence of giant extreme waves exceeding 20 m in height has been recognized as a severe problem around the 1990s. However, a comprehensive understanding and prediction of extreme waves has been a difficult task. One of the challenges of tackling this intriguing phenomenon was to quantify the role of nonlinear focusing in accelerating the occurrence probability of extreme waves. The studies became one of the main research topics in the dynamics of ocean waves, and extensive knowledge has been acquired until today. The research activities on extreme waves derive from applied mathematics, physics, physical oceanography, and engineering. To date, no summarized textbook devotes itself to describing an overview of extreme waves covering the science of extreme waves, prediction, and their engineering applications.

This book aims to provide a holistic view of extreme ocean waves for scientific and engineering applications. We also aim to educate ocean and coastal engineers on the fundamental theory of extreme waves to design structures, ships, offshore wind farms, and other marine platforms. The first two chapters review the definitions of extreme waves and the history of their study to date. Chapters 3 and 4 summarize the observations of extreme waves. Chapters 5–7 describe the theory and modeling of extreme waves in various sea-state configurations. Chapters 8 and 9 introduce extreme wave prediction systems and their consequent impact. Finally, application examples of extreme wave research and their applications to various engineering design processes are presented in Chapters 10–13.

The book's goal is not to answer all outstanding questions on freak/rogue waves. Instead, we intend to provide a comprehensive overview of the current status of our understanding of freak/rogue waves and to show how those are implemented in engineering practices.

Nobuhito Mori
Takuji Waseda
Amin Chabchoub

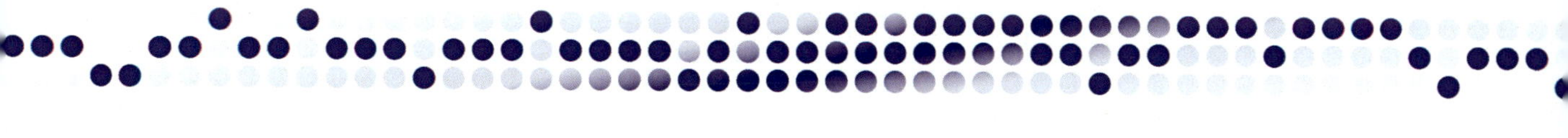

1

What is a rogue/freak wave?

Nobuhito Mori[1], *Takuji Waseda*[2] *and Amin Chabchoub*[1,3]
[1]Disaster Prevention Research Institute, Kyoto University, Japan [2]Graduate School of Frontier Sciences, The University of Tokyo, Japan [3]School of Civil Engineering, The University of Sydney, Sydney, NSW, Australia

Introduction

The term *rogue wave* or *freak wave* has long been used by maritime, ocean engineering, and coastal engineering professionals to describe waves that are much higher than would be expected based on surrounding sea conditions (e.g., Draper, 1966; Dysthe et al., 2008; Mori, 2019). It was initially referred to as a *freak wave*, but the term *rogue wave* has gained more popularity since the 2000s. Hereafter, such extreme waves in the ocean will be referred to as *rogue waves* in this chapter. Both terminologies are generally used together. Therefore, it is up to the authors of the chapters to decide whether to use *rogue wave* or *freak wave* following contents.

Numerous encounters with anomalous waves have been reported in the history of the ocean. Evidence of such anomalous waves has become increasingly substantial since the 1980s. Klinting and Sand (1987) reported some of the extreme wave events recorded in the North Sea, the so-called rogue/freak waves. The basic definition of a rogue wave is a wave with a height that is more than twice the height of a significant wave.

$$H_{\max} \geq 2H_{1/3} \tag{1.1}$$

where $H_{\max}$ is the maximum wave height of the wave train, and $H_{1/3}$ is the significant wave height, defined as the highest one-third of waves within a wave train.

Klinting and Sand (1987) provided a more detailed distinct definition of the rogue wave, consisting of three rules (denotes KS criteria).

1. Wave height exceeds twice the significant wave height as described in Eq. (1.1).

Science and Engineering of Freak Waves.
DOI: https://doi.org/10.1016/B978-0-323-91736-0.00006-7

2. Wave heights are more than twice as high as previous/following wave heights.
3. Wave crest height is more than 65% of its wave height.

If a wave satisfies all three criteria, the shape of a single outstanding wave can be considered as rogue within the wave train. However, as research on rogue waves focused on extreme waves, the KS criteria have been somewhat relaxed, and only the first condition, that is, $H_{max} \geqq 2H_{1/3}$, is generally used to define rogue waves.

In the first half of the 1990s and with the first Draupner recording, as will be elucidated in the next Section, there has been intensive discussion in the engineering community. It was reported that the cause of the occurrence of freak waves in the open ocean in general was largely due to third-order or higher-order nonlinear interference at about the same time, according to numerical simulations and tank experiments. Since around 2000, the research on rogue wave began to attract attention from both science and engineering fields, and several interdisciplinary international conferences have been held (e.g., Olagnon and Athanassoulis, 2001).

The progress of research in the past two decades has been particularly significant. More recently, the topic has also attracted the interest of wave physicists. Particularly, the optics community has initiated drawing powerful interdisciplinary analogies of the rogue wave phenomena in different wave systems (Dudley et al., 2019). Below a summary of the historical evidence and causes of freak waves.

Ocean rogue wave observations

Sailors have observed, remembered, and reported extreme wave encounters (Draper, 1966). For example, the captain of the cargo ship "Junior" reported waves estimated to be 30.5 m (100 ft) high. A reliable and well-known report states that the waves encountered by the US warship "Ramapo" in the North Pacific in 1993 were estimated to be 34.1 m (112 ft) high. There are numerous accounts of such waves striking passenger, container ships, oil tankers, fishing vessels, and offshore and coastal structures, sometimes with devastating results; it is believed that more than 22 supercarriers were lost to rogue waves between 1969 and 1994 (Fig. 1.1).

Three well-studied examples and new data are the Draupner "New Year Wave" in the North Sea, winter storm waves in the Japan Sea, and recently observed typhoon-generated waves, as shown in Fig. 1.2. The Draupner wave observed in the North Sea in January 1995 is shown in Fig. 1.2A. It was recorded by a laser instrument installed offshore. This record of the Draupner wave is widely discussed in the scientific literature. The wave height is 25.6 m with

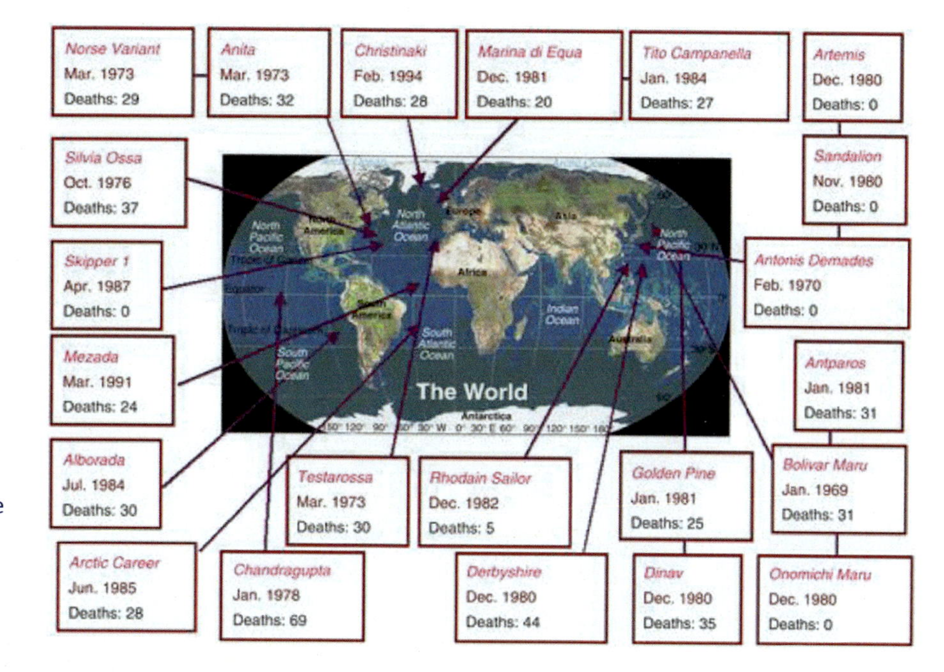

Figure 1.1 Locations of supercarriers were assumed to have been lost by rogue waves between 1969 and 1994. *© C. Kharif and E. Pelinovsky. used by permission.*

$H_s = 11.9$ m. The other typical profile of winter storm-generated rogue waves recorded in the Japan Sea is also shown in Fig. 1.2B (Mori et al., 2002). This dataset includes 10 rogue waves from 4 years of continuous measurements by a laser instrument mounted on the platform during the winter season. A few of the measured rogue waves fully meet the KS criteria. Based on the wavelet spectra, it appears that the energy density is carried over to the high-frequency component at the rogue wave moment for the case shown in Fig. 1.2B. However, the Draupner rogue wave time series show smooth grouping and less energy at high frequencies. It should be noted, however, that for most extreme waves, the significant wave heights are only 2 to 4 m high.

Fig. 1.2C shows the wave profile of the drifted buoy observed near the typhoon eye wall in the middle of the Western North Pacific in 2021. This data is one of many rogue waves as a part of a large field campaign using drifted buoy network (Fig. 1.2C shows one of the extreme cases). Many different shapes of rogue waves were observed in the typhoon wave condition, with significant wave heights exceeding 5 m.

Some of the observed rogue waves are single outstanding waves in wave trains that meet all of the KS criteria, while others are not. It indicates that the rogue wave formation consists of a convergence of linear and nonlinear waves. This will be discussed later in this chapter.

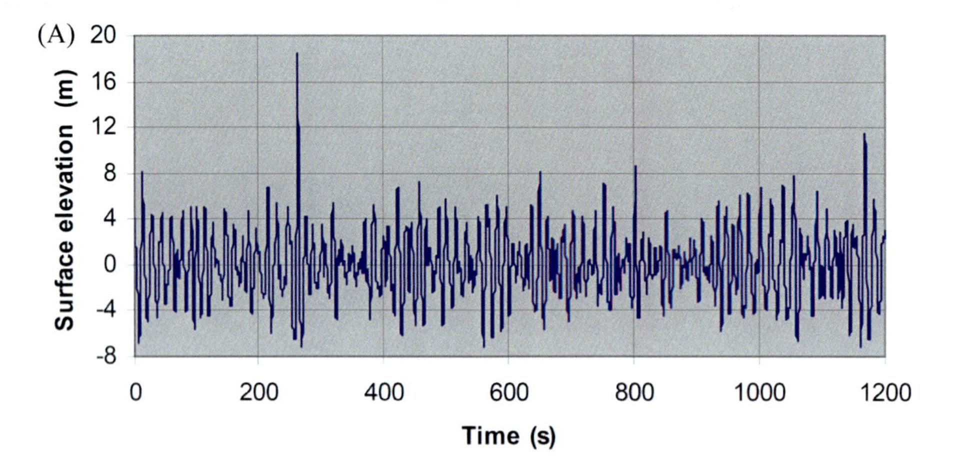

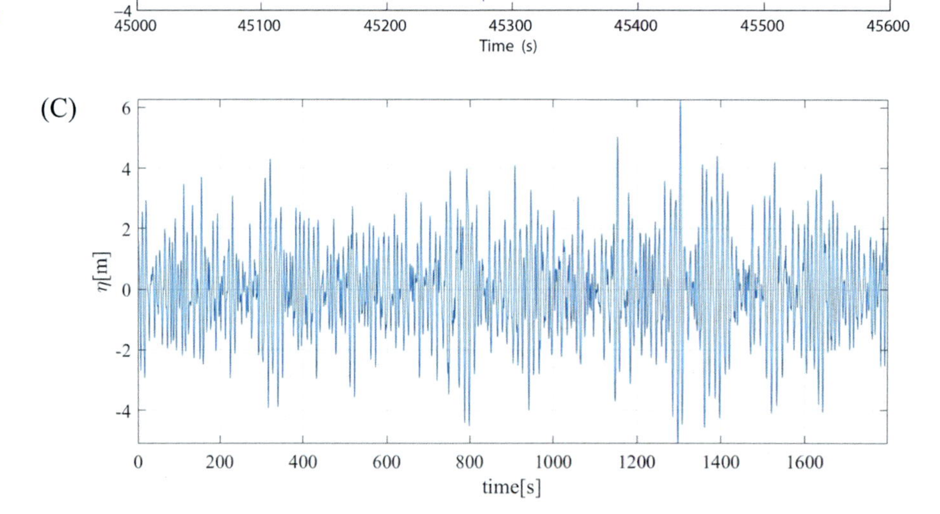

Figure 1.2
Three examples of rogue waves. (A) The Draupner wave was recorded on January 1, 1995, at the Draupner platform in the North Sea (Haver, 2004). (B) The Yura wave is a typical single outstanding wave recorded at the Yura platform in the Sea of Japan in 1987 (Mori et al., 2002). (C) The Pacific typhoon wave occurred in the middle of the Western North Pacific on July 19, 2021. *Panel (A) is reprinted with permission from the Proceedings of Rouge Wave Workshop 2004. Panel (B) is reprinted with permission from Ocean Engineering, Volume 29 © 2002 by Elsevier.*

Wave evolution in the ocean

Several mechanisms cause rogue waves in the ocean. It is important to understand these mechanisms and to estimate how large the rogue waves are and how often they occur. Surface waves are generated by wind and grow in time and space according to storm characteristics. As the wind continues to blow, the wave energy grows, and characteristic frequency shifts to lower frequencies. The evolution of wave spectra describes such a physical process of wave evolution (Mitsuyasu, 1970, and Hasselmann et al., 1973).

The wave spectrum is generally described by frequency and direction or two-dimensional wave number space. The evolution of the wave spectrum is driven by energy input from the wind, energy transfer from high to low frequencies due to nonlinear wave–wave interactions, and dissipation due to breaking waves. Energy-based numerical wind wave models (wave spectral models) simulate the development of wave spectra and are commonly used for wave prediction and hindcasts. However, the spectral wave models cannot handle individual wave profiles because they describe averaged wave energy in frequency and direction. Therefore, the spectral wave models cannot directly provide information on rogue waves.

A phase-resolving model e.g., Zakharov equation in deep water (Zakharov, 1967), Boussinesq equation in shallow water (Dingemans, 1997) described kinematic and dynamic boundary conditions of water surface waves directly. Consequently, the phase-resolving model can handle changes in individual wave profiles accurately in comparison with the wave spectral model. On the other hand, the phase-resolving model cannot include the effects of wind energy input and related dissipation due to wave breaking consistently with the observation data as similar to the spectral wave models. From a general computational fluid dynamics point of view, there is also a direct simulation possibility using the Navier–Stokes equations. The direct simulation is quite accurate and can include wind energy input and dissipation by wave breaking. However, it is difficult to simulate for longtime evolution of random wave fields. As such, the Navier–Stokes equation is mostly used to evaluate the wave force of rogue waves acting on ships and structures, as will be discussed in Chapters 10 to 12.

Generally, the spectral wave model is used for the evaluation of waves in a wide area where the development of wind waves can be taken into account, and the phase-resolving model is used for evaluation in a narrow region where there is no wave energy change. The gap between the two models is a limitation in the study of rogue waves.

Causes of rogue waves

Rogue waves represent a highly localized concentration of wave energy within a wave group. The most important point is the mechanism of wave-focusing, which can concentrate energy at a specific place and time.

Wave-focusing mechanisms can be classified into four types. The first is linear focusing, and the second is nonlinear focusing. These are self-focusing mechanisms that are independent of the surrounding ocean environment. The other mechanisms are wave–current interactions and bathymetry focusing. These are external-focusing mechanisms due to the surrounding ocean environment. This section provides an overview of these four mechanisms.

Linear wave focusing

Deep water gravity waves are dispersive depending on the wavelength. This is because the phase and group velocity are inversely proportional to frequency (i.e., long waves propagate faster than short waves). As the wind waves have a finite spectrum bandwidth range, the waves consist of different amplitudes and frequencies depending on the spectra shape.

If wave nonlinearity is relatively small ($ak << 1$; where a is wave amplitude, and k is wave number), a random wave train can be represented by a linear superposition of periodic waves. Due to a random combination of different wave number components, an extremely localized wave, a rogue wave, can be generated by the dispersive linear wave focusing. Therefore, linear wave focusing can occur under general conditions. For example, if a random wave train in one-dimensional space can assume stationary, linear, and narrow-band spectrum conditions, the probability density function of wave height H can be described as

$$p(H)dH = \frac{1}{4}He^{-\frac{1}{8}H^2}dH, \tag{1.2}$$

where wave height H is normalized by the root mean square (rms) value of water surface elevation η_{rms}. Eq. (1.2) is the so-called Rayleigh distribution, and it represents the probability of wave height in the random wave train. Based on Eq. (1.2), the distribution of maximum wave height H_{max} can be expressed as

$$p_{\mathrm{m}}(H_{\mathrm{max}})dH_{\mathrm{max}} = \frac{N}{4}H_{\mathrm{max}}\xi\exp(-N\xi)dH_{\mathrm{max}} \tag{1.3}$$

$$\xi = e^{-\frac{H_{\mathrm{max}}^2}{8}} \tag{1.4}$$

where N is the number of waves in a train. From Eq. (1.3), the distribution of maximum wave heights is given by the number of waves N.

Considering KS criteria by Eq. (1.1), the rogue wave condition is a maximum wave in a wave train that is greater than twice the significant wave height (it corresponds to $H_{\mathrm{max}}/\eta_{\mathrm{rms}} \geq 8$). The occurrence probability of rogue waves can be obtained.

$$P_{\mathrm{freak}} = 1 - \exp\left[-e^{-8}N\right] \tag{1.5}$$

Eq. (1.5) presents the occurrence probability of rogue wave for a linear random wave train. For the case of $N = 100$ and 1000, the occurrence probability of a rogue wave predicted by Eq. (1.5) is 3.3% and 28.5%, respectively (e.g., Mori and Janssen, 2006). Assuming $T_{1/3} = 10$ seconds, the number of waves $N = 1000$ corresponds to 3 hours storm. A rogue wave appears with a reasonable probability, even in a linear wave field.

A linear wave focusing can be the primary rogue wave mechanism in most parts of the ocean. That said, the second-order wave nonlinearity only modifies the wave profile in a form of asymmetry between crests and troughs (crests are sharper and troughs are flatter), but does not change in Eqs. (1.3) or (1.5). It is because the second-order wave nonlinearity does not transfer energy between spectral components. However, higher-order models account for weak resonant interactions between these wave components and enhance the occurrence frequency of rogue waves.

Nonlinear wave focusing

The second mechanism of the cause of extreme waves results from wave interactions by third-order nonlinearity, which is the underlying mechanism of the four-wave quasi-resonant interaction process (Janssen, 2003). Numerical calculations have shown that four-wave interactions with third-order nonlinear interactions in the deep ocean can generate rogue waves with a single steep crest. The theoretical background of rogue wave generation was clarified by Janssen (2003).

The second-order nonlinearity plays a wave asymmetry concerning the short-time evolution of the wave train, $O(\omega^{-1}\varepsilon^{-2})$, where ω is an angular frequency, and $\varepsilon = ak$ is wave steepness. There is a change of spectral shape by the second-order nonlinear interactions. However, resonant interactions due to third-order nonlinearities allow energy exchange between wave components and trigger nonlinear focusing of extreme wave generation. The timescale of resonant four-wave interactions is longer than second order, $O(\omega^{-1}\varepsilon^{-4})$, and two orders of magnitude are longer than second-order nonlinearities. However, quasi-quadruple-wave interactions with $O(\omega^{-1}\varepsilon^{-2})$ time scales play an important role in the generation of extreme waves, as shown below.

Modulational instability is a quasi-resonant interaction process; that is, the wave number and frequency satisfy the following conditions

$$\vec{k_1} + \vec{k_2} + \vec{k_3} + \vec{k_4} = 0 \tag{1.6}$$

$$\vec{\omega_1} + \vec{\omega_2} + \vec{\omega_3} + \vec{\omega_4} \leq \varepsilon^2 \tag{1.7}$$

In particular, modulation instability occurs when the two wave numbers are the same, $\vec{k_1} = \vec{k_2}$ and $\vec{k_3}$ and $\vec{k_4}$ are two sidebands separated from $\vec{k_1}$ by Δk, which should be small to satisfy the conditions of Eqs. (1.6) and (1.7). Standard equations of motion describing the evolution of the wave spectrum for large times and exact resonances have been established. However, an extension to quasi-resonant interactions allows energy exchange on shorter time scales to excite energy transfer near the peak frequency. Quasi-resonant interactions change the surface properties from a Gaussian state. The Benjamin–Feir index (BFI) by Janssen (2003) describes the conditions for quasi-resonant interactions for unidirectional wave trains.

$$\mathrm{BFI} = \frac{\sqrt{2}\varepsilon}{\delta_\omega} \tag{1.8}$$

where $\delta_\omega = \sigma_\omega/\omega$ is the relative spectral frequency bandwidth, and σ_ω is the characteristic spectral bandwidth near the peak frequency. When BFI is sufficiently large, the spectrum and the fourth-order moment of the surface elevation change because of quasi-resonant interaction. If BFI is sufficiently large, the occurrence probability of a rogue wave is larger than a linear random wave, as shown in Eq. (1.5).

Furthermore, wave directionality can reduce such nonlinear enhancement of rogue waves (Mori et al., 2011; Waseda et al., 2009). The details of nonlinear focusing will be discussed in the following chapters.

Bathymetry focusing

Bathymetry focusing can be achieved by refracting waves by variable bottom topography and currents, essentially following a linear wave mechanism. As the waves propagate into shallower water and their wavelength is equal to the depth of the water, they are refracted, and their crests follow the topography and become steeper because of the conservation of energy. Thus, on irregular shorelines, wave energy may be concentrated in certain locations, resulting in anomalous waves. However, such focusing is easily predictable because extreme waves are always generated around the same location. We refer to Chapter 7 for more details on this focusing mechanism, which also include the crucial role of wave nonlinearity in the exact approximation of the focusing location.

Wave–current interaction

The simplest example, which can be considered to illustrate the mechanism, is a wave propagating from still water into an incoming flow. Waves with phase velocity c_p in still water are completely blocked by the counter-propagating flow of only $c_p/4$. Extreme waves of $c_p \sim 15\ \mathrm{m\ s^{-1}}$ do not stop but are delayed and shortened in wavelength as they propagate against the flow. The wave flux (energy/natural frequency) is conserved, meaning that the currents give energy when they interact with the waves.

The situation can be more severe for strong currents such as the Kuroshio Current in the Northwest Pacific, the Gulf Stream in the Northeast Atlantic, the East Australian Current, and the Agulhas Current in South Africa. When a wind wave or swell travels in the opposite direction to the current, it can become trapped where the current spreads or meanders in ray tracings. The trapped waves are refracted toward the center of the current and form an oscillatory path due to reflections (or caustics) near the edges. The wave amplitude is significantly amplified during reflection compared to its value at the current center. If waves from adjacent rays around the reflection are constructively added, rogue waves may be generated. When wave groups propagate in opposing uniform currents, strong non-Gaussian properties of the wave field are expected, whereas waves are more stable in the presence of shear currents (Thomas et al., 2012).

Summary

A common mechanism for the generation of rogue waves in the open ocean is the convergence of linear and nonlinear waves; the quasi-resonant interaction of the four waves plays an important role in determining the statistical properties of surface waves. The historical background of rogue waves will be summarized in Chapter 2.

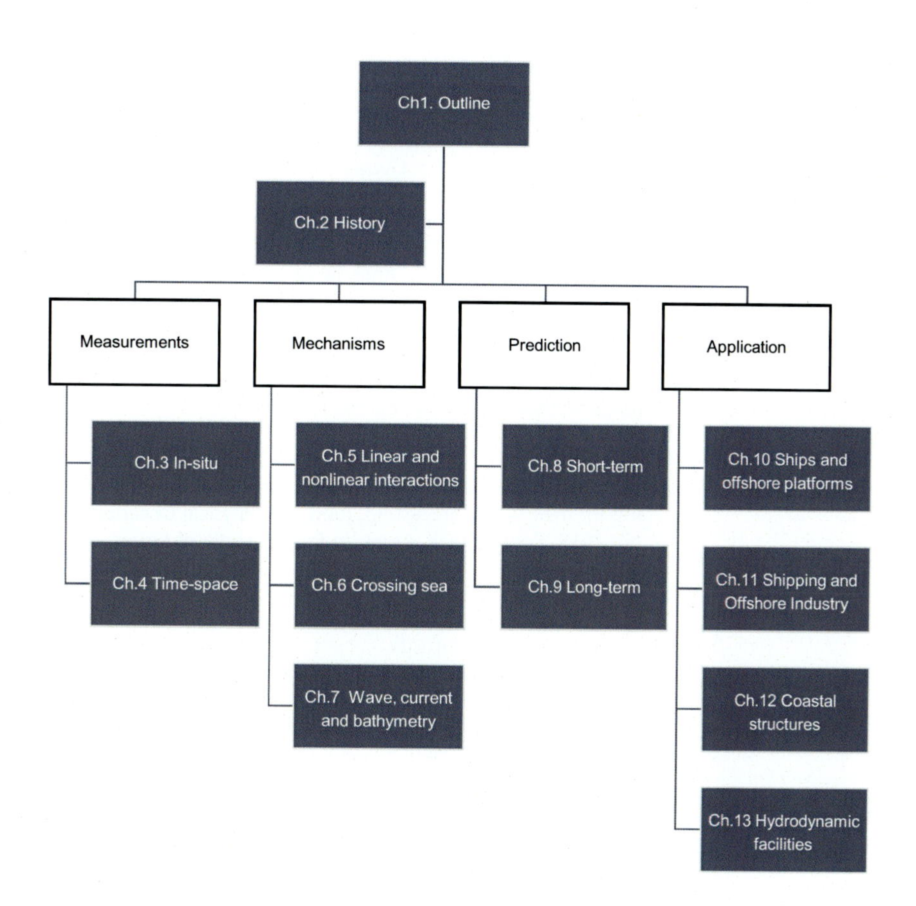

Figure 1.3 Outline of this textbook.

Indeed, there are several essential debates on rogue waves. First, an important topic is the measurement of rogue waves in the ocean. This measurement is the first step toward understanding of the phenomenon (see Chapters 3 and 4). The second important topic is the modeling and understanding of rogue wave physics (see the end of this Chapter and Chapters 5 to 7). The third important theme is the prediction of short- and long-term rogue wave dynamics (see Chapters 8 and 9). In addition, applying current knowledge of rogue waves (or extreme wave modeling) to engineering problems is particularly important for scientific, engineering, and industrial problems. Therefore, it will be emphasized in the last part of this book (see Chapters 10 to 13). Fig. 1.3 is a visual summary of the chapters of this book.

References

Dingemans, M.W., 1997. Wave propagation over uneven bottoms- Part 2. Advanced Series on Ocean Engineering 13, 473–688.

Draper, L., 1966. 'Freak'ocean waves. Weather 21 (1), 2–4.

Dudley, J.M., Genty, G., Mussot, A., Chabchoub, A., Dias, F., 2019. Rogue waves and analogies in optics and oceanography. Nature Reviews Physics 1 (11), 675–689.

Dysthe, K., Krogstad, H.E., Müller, P., 2008. Oceanic rogue waves. Annual review of fluid mechanics 40, 287–310.

Hasselmann, K., Barnett, T.P., Bouws, E., Carlson, H., Cartwright, D.E., Enke, K., et al., 1973. Measurements of wind-wave growth and swell decay during the Joint North Sea Wave Project (JONSWAP). *Ergaenzungsheft zur Deutschen Hydrographischen Zeitschrift*. Reihe A .

Haver, S., 2004. A possible freak wave event measured at the Draupner Jacket January 1 1995. *Rogue waves*. Ifremer, Brest, France.

Janssen, P.A., 2003. Nonlinear four-wave interactions and freak waves. Journal of Physical Oceanography 33 (4), 863–884.

Klinting, P., Sand, S.E., 1987. Analysis of prototype freak waves (No. NEI-DK-190, CONF-8706402-1).

Mitsuyasu, H., 1970. On the growth of the spectrum of wind-generated waves. Coastal Engineering in Japan 13 (1), 1–14.

Mori, N., Janssen, P.A., 2006. On kurtosis and occurrence probability of freak waves. Journal of Physical Oceanography 36 (7), 1471–1483.

Mori, N., Liu, P.C., Yasuda, T., 2002. Analysis of freak wave measurements in the Sea of Japan. Ocean Engineering 29 (11), 1399–1414.

Mori, N., Onorato, M., Janssen, P.A., 2011. On the estimation of the kurtosis in directional sea states for freak wave forecasting. Journal of Physical Oceanography 41 (8), 1484–1497.

Mori, N., 2019. Rogue/Freak waves, In Encyclopedia of Ocean, Third EditionSciences, Volume 3. pp. 642–649.

Olagnon, M., Athanassoulis, G.A., 2001. Rogue Waves 2000: Proceedings of a Workshop, Organized by Ifremer and Held in Brest, France, 29–30 November 2000, Within the Brest SeaTechWeek 2000 (Vol. 32).

Thomas, R., Kharif, C., Manna, M., 2012. A nonlinear Schrödinger equation for water waves on finite depth with constant vorticity. Physics of Fluids 24 (12), 127102.

Waseda, T., Kinoshita, T., Tamura, H., 2009. Evolution of a random directional wave and freak wave occurrence. Journal of Physical Oceanography 39 (3), 621–639.

Zakharov, V.E., 1967. The instability of waves in nonlinear dispersive media. Sov. Phys. JETP 24 (4), 740–744.

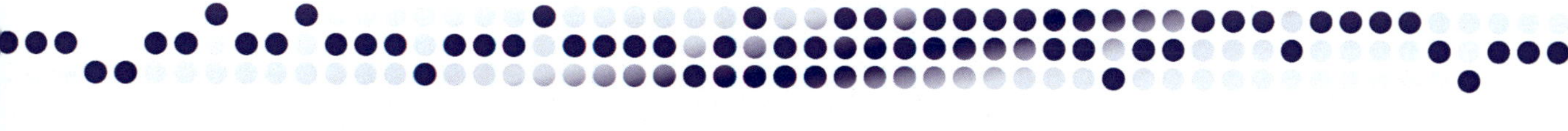

History of freak/rogue wave research

Takuji Waseda
Graduate School of Frontier Sciences, The University of Tokyo, Japan

Introduction

I am on a research vessel navigating through the *Furious 50 s* observing waves from the bridge. The wind is over 40 knots; the sea surface is covered by white streaks of foam caused by energetic breaking waves. The ship is rolling and pitching and, every once in a while, dives into the wave trough, creating a huge splash that reaches the bridge windshield (Fig. 2.1). Sometimes, slamming is followed by a quake of the ship. Now, as I look ahead of the ship, I see waves propagating toward the ship. Occasionally, I spot a large wave with a relatively long wave crest propagating from the head sea. The wave gradually increases its amplitude, and the crest overturns and plunges into the wavefront leaving a streak of white foam. The second wave, slightly lower in amplitude, propagates behind that breaking wave. This second wave gradually gains amplitude and eventually breaks. Then, the following wave grows and breaks. The sequence seems to continue. On the other hand, I see a more pyramidal breaking wave as a result of a collision of two waves propagating at an angle, somewhat resembling the sloshing waves in an enclosed basin. Such breaking waves occur more spontaneously without a precursor; it is unpredictable.

So, where is the freak wave? Now I switch my attention to find a freak wave. From the textbook, we learn that a freak wave occur every 3000 waves. For the wave period of 10 seconds, I need about 8 hours to find one. That is a long time. But, I have been on the bridge for many hours by now, so one of the waves I saw may have been a freak wave. How can I tell? Another thing we learned about freak wave is that it is much larger than the surrounding waves. Well, that depiction somewhat resembles that of the beautiful energetic breaking wave I saw. But is a freak wave a breaking wave?

After all these years of studying ocean waves, I feel that I have not learned enough. If I do not understand ocean waves, how can I explain what a freak

Science and Engineering of Freak Waves.
DOI: https://doi.org/10.1016/B978-0-323-91736-0.00013-4

Figure 2.1 Images from the bridge of an icebreaker Shirase navigating in the *Furious 50 s. Photographs by T. Waseda.*

wave is? The explanation of freak waves that comes to my mind is the typical time series with one outstanding wave, the enhanced tail of the statistical distribution, the nonlinear Schrödinger equation, etc. Are they useful or not? Are they used in the design of an offshore platform? What about the design of a ship? Can we predict freak waves? Do we know what freak waves look like? If I collect the existing knowledge, maybe I can understand what a freak wave is. So, that is my motivation to write about the history of freak waves, as a prelude to the following chapters that present the state of the art of freak wave research.

The origin of its name

A freak wave, a very unusual wave, is also named a rogue wave, a dangerous wave that does not belong to the rest of its group. The wave is also called a mad-dog wave in Taiwan, a wave that suddenly bites. These terminologies imply that freak/rogue waves are outliers of a given population. Thus the origin of these naming is that they are statistical anomalies. The occurrence of the wave is rare. In space, it appears as an isolated high wave, and in time, it seems to appear from nowhere (Akhmediev et al., 2009).

The wave is called a "Sankaku Nami" or triangular wave in Japan. This is the only name given to the "freak/rogue wave" that represents geometry. It is debatable whether the freak/rogue wave is long-crested or short-crested, but the "Sankaku Nami" represents a breaking wave in Japanese literature (e.g., Kaniko-sen by T. Kobayashi). So, most likely the seafarer in Japan uses the word "Sankaku Nami" to represent a steep wave that is close to or breaking. The artistic depiction of a steep breaking wave known as the Great Wave off Kanagawa by Hokusai may have influenced the naming.

Another important name is the "Three Sisters." Instead of a single wave that stands out from the surrounding waves, the Three Sisters suggest that the highest wave is accompanied by two waves in the front and in the rear that are comparable to the largest wave. The three waves propagate together, and if one observes for an extended time, one will see that the largest wave gradually reduces its amplitude while the succeeding wave gains energy and replaces the initial large wave. Another wave appears from behind and becomes the third wave. A coherent structure of three waves is maintained. This is a consequence of the energy propagation of surface gravity waves at the group velocity which is half of the phase speed. The seemingly coherent structure is called the wave group (Donelan et al., 1972). When the waves break in a group, the breakers mark strong signals in a marine radar coherently aligned along the group speed line while each signal propagates along the phase speed line (Tulin, 1996).

In this textbook, we use both the terms freak waves and rogue waves. Both are commonly used in scientific and engineering communities. The term rogue wave is used extensively in the field of optics since the first discovery of optical rogue waves in 2007 (Soli et al., 2007). Since then, rogue waves tend to be used more commonly in the field of hydrodynamics as well, despite freak wave was used more at the beginning when it was discovered. It is interesting to note that soliton was also first discovered in shallow water and now it appears in a variety of fields without links to the hydrodynamic discovery.

Weak nonlinear process of the evolution of ocean waves

Ocean waves grow under the action of wind, and as the wave height increases, so does the wavelength. This fetch law is the most fundamental law that governs the growing wind sea. The increase of the wavelength is equivalent to the decrease of the peak frequency of the spectrum of ocean waves and is called spectral downshifting. To date, the most commonly agreed mechanism of spectral downshifting is the nonlinear transfer of energy first formulated by Hasselmann (1962). Surface gravity waves interact with each other and exchange energy when the resonant condition of the four waves is satisfied (Phillips, 1960). Hasselmann considered interaction among all the possible quartet combinations of a directional wave field and showed that the net result is to transfer energy from high frequency to low frequency of the spectrum. This is a conservative process, and the rate of transfer depends strongly on the spectral shape. Under the balance of wind input, dissipation, and nonlinear energy transfer, ocean wave spectra maintain their self-similar shape represented by the equilibrium tail proportional to ω^{-4} (Toba, 1973) or ω^{-5} (Phillips, 1958). Around the peak, the spectral energy is enhanced, and its shape can be expressed by a narrow Gaussian function. The JONSWAP spectrum (Hasselmann et al., 1973) takes into consideration both the equilibrium tail (ω^{-5}) and the peak enhancement and is widely used in scientific and engineering studies.

Zakharov (1968) derived a deterministic evolution equation for the complex amplitudes of the wave components of a directional wave field. This short paper contains several important findings that will be exploited in the next half-decade. We can derive Hasselmann's integral under the assumption of a quasi-Gaussian closure (e.g., Krasitskii, 1994). In the narrowband approximation, the nonlinear Schrödinger equation (NLS) can be derived from Zakharov's equation (e.g., Stiassnie, 1984). The use of NLS is immediately followed by its application to study the instability of a wave train. The stability of a monochromatic wave is studied as a special case by Zakharov (1967, 1968). He shows that the wave is unstable to a pair of waves only when "the effects of nonlinearity are less than the effects of dispersion" which is satisfied by the surface gravity waves. Benjamin and Feir (1967) showed experimentally and theoretically that the Stokes wave is unstable to a pair of sideband waves whose condition is that $0 < \frac{\delta\omega}{\omega} \leq \sqrt{2}ak$, where $\delta\omega$ is the frequency difference of the pair of sidebands from the carrier wave ω, and ak is the steepness of the carrier wave (a is the amplitude, and k is the wave number).

Several studies were conducted to derive an evolution equation for the statistical properties of ocean waves. A common approach was to start with deriving an equation for a spectral correlation function from Zakharov's equation. The complicated evolution equation greatly simplifies when a quasi-Gaussian closure is assumed or equivalently assuming that the wave field is homogeneous all the time (Yuen & Lake, 1982). Thereby, Hasselmann's equation is derived from Zakharov's equation. By allowing inhomogeneity of the wave field, Janssen (2003) extended the theory to explain how the statistical properties change in time for a given spectrum. Janssen's study suggests that in the limit of a short timescale, the fourth-order cumulant or the kurtosis changes in time. The evolution includes both exact and nonresonant (or quasi-resonant) interactions. In the physical space, the increase of kurtosis corresponds to the increase in the occurrence probability of a freak wave. This result was experimentally demonstrated by Onorato et al. (2004). The deviation from the Gaussian state can be directly related to the modification of the probability density function through the Gram–Charlier expansion (Mori & Janssen, 2006). More details are provided in Chapter 5 of this book.

A further extension was made to derive a generalized kinetic equation (GKE) that is free of the quasi-stationarity assumption (Annenkov & Shrira, 2006). Allowing for a near-resonant interaction at play, they demonstrate that exclusion of the exact resonant interaction has a small effect on the overall evolution. This result is consistent with the earlier study by Tanaka (2001) employing a direct numerical simulation of a discrete wave system. Tanaka demonstrated that even as short as 25 wave periods, the Hasselmann-like nonlinear energy transfer was achieved for the given directional wave field. The GKE was rederived by Gramstad and Stiassnie (2013) and further implemented in an operational wave model (Gramstad & Babanin, 2016). However, no noticeable change in the kurtosis was found for a realistic directional wave field.

As of now, most studies consider that weak nonlinear process plays a significant role in forming the directional spectra of ocean waves. However, when it comes to the generation of a freak wave, sometimes weak nonlinearity is disregarded, and the linear process is considered. If the linear process is dominant in the ocean, how can we explain spectral downshifting? Donelan et al. (2012) presented a linear wave model but with an inclusion of a dissipation-driven downshifting mechanism. They consider that the energy is redistributed by a spilling breaker. Donelan and Magnusson (2017) further extended their idea to explain how the linear superposition of a wave group can explain the formation of a freak wave. Therefore, Donelan provides a linear theory that explains both the spectral downshifting of directional wind sea and the formation of a freak wave. The caveat is the coherence of a wave group requiring weak nonlinearity.

Marine accidents and reproductions of observed wave fields

Ship incidents were studied considering possible encounters with a freak wave. Toffoli et al. (2005) studied 270 incidents and showed, e.g., that crossing seas can be a possibly dangerous condition (see Chapter 6 for a comprehensive review). However, there was no indication that the ships encountered a freak wave as the steepness was in general small. On the other hand, Waseda et al. (2012) investigated seven incidents near Japan and found that five cases occurred when the directional spectrum was the narrowest within a day or so. Among those, two cases were studied extensively. The first case is the Onomichi-Maru incident that occurred near Japan in 1980. Marine accident inquiry reported that there were two wave systems that produced a crossing sea state. At the time of the incident, the media used the term "Sankaku Nami" extensively. However, the reanalyzed wave field indicated that the spectrum was unimodal and a wind sea grew under a running fetch condition with not much downshifting (Waseda et al., 2014). The second case is a crossing sea condition where a swell from the Baiu front mixed with the wind sea under atmospheric low (Suwa-Maru incident: Tamura et al., 2009). The swell grew at the expense of wind sea energy. Both studies indicate that the spectrum narrowed due to meteorological forcing.

Following these works making use of a directional spectrum to infer the possible occurrence of freak waves, several studies were conducted using a phase-resolved wave model to reproduce high-order statistics as well as the possible freak wave shape (Bitner-Gregersen et al., 2014; Dias et al., 2015; Fedele et al., 2016; Trulsen et al., 2015; Fujimoto et al., 2019). A Monte Carlo simulation based on the high-order spectral method is commonly used in all these studies. All the studies showed an increase in excess Kurtosis, but the values were a lot less than what the experimental works have shown (Waseda et al., 2009; Onorato et al., 2009). Some studies showed that the enhanced probability density of the crest height can be explained by the second-order theory

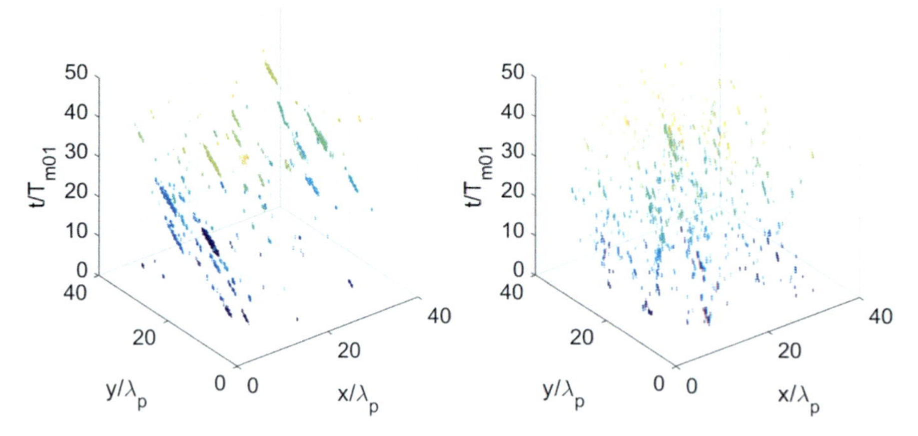

Figure 2.2 Freak wave groups are simulated by a high-order spectral model. The left diagram shows the case for a narrow and unimodal directional spectrum. The right diagram shows the case of a broad and bimodal directional spectrum. Left shows a mixture of elongated (long life) and spot-like (short life) freak wave groups. *Adopted from Fujimoto, W., Waseda, T., Webb, A. (2019). Impact of the four-wave quasi-resonance on freak wave shapes in the ocean. Ocean Dynamics, 69(1), 101–121*

(e.g., Tayfun and Fedele, 2007). On the other hand, the lifetime of the freak wave group extends as the directional spectrum narrows (Fujimoto et al., 2019). The implication is that random focusing freak waves and nonlinear focusing freak waves coexist, and their population depends on the directional spectrum.

The crux is to infer the generation mechanism of freak/rogue waves from limited information. A recent observational study showed that the linear superposition of waves may explain the observed rogue wave (Gemmrich and Cicon, 2022). They showed that the phases of each wave component match when the extreme wave is observed. However, a recent study on modulated wave trains also demonstrated a phase convergence as the wave energy self-focuses (Houtani et al., 2022). Thus from a single-point measurement, it is not possible to identify the generation mechanism. For that reason, instead of a single-point measurement, spatiotemporal data are analyzed (Benetazzo et al., 2015, and Chapter 4 for a thorough explanation). As aforementioned, when the nonlinear effect becomes important, the lifetime of freak waves extends (Fig. 2.2, Fujimoto et al., 2019). The coherent wave group propagates and therefore should be identified if a spatiotemporal observation is available (e.g., Waseda et al., 2021).

Impact of freak waves on offshore structures and ships

Regardless of the generation mechanism, steep and high waves will have a large impact on ships and offshore structures. In the 1990s, numerous studies were conducted to understand the causes of the ringing of offshore tension leg platforms (TLPs) and gravity-based structures (GBSs). Ringing is a high-frequency transient response of TLPs and GBSs at around 2 to 4 seconds which is much shorter than the typical incident wave period. In a controlled laboratory experiment, both the random linear focusing waves (e.g., Chaplin et al., 1997) and the modulated wave trains (e.g., Levi et al., 1998) were used. In both cases, the waves were not breaking but were close to breaking and, therefore, strongly nonlinear. Welch et al. (1999) conducted experiments with breaking waves and showed that the local steepness was a relevant factor. These studies considered the nonlinearity of the incident waves instead of considering high-order diffraction forces due to Stokes waves.

The impacts of extreme waves on ships have been studied mostly in accord with ship accidents. Among the ship accidents summarized in Kharif and Pelinovsky (2003), the Onomichi-Maru incident that occurred in 1980 has been studied extensively from a viewpoint of ship design (e.g., Waseda et al., 2014). The shocking image taken after the loss of a bow section implied that there is a high-frequency bending mode of the ship triggered by the encounter of ships and the extreme waves. This somewhat resembles the impacts of extreme waves on offshore structures. In the ship accident case, slamming impact triggers whipping of the ship hull, and the bending moment increases (Houtani et al. 2019, Chapter 10 of this book). The question is what kind of wave is most dangerous for the ship?

In a laboratory tank, several techniques have been employed to generate isolated extreme waves. Linear dispersive focusing is a classic method to generate a physically isolated high wave with a broad spectrum. Linear focusing wave has been used to study linear ship responses to incoming waves. As the amplitude becomes higher, local nonlinearity increases and several studies showed that the four-wave resonant interaction may play a significant role (Barratt et al., 2020). Wave trains can be generated in two ways considering their nonlinear evolution. First, under a narrow spectral approximation, the amplitude of a monochromatic wave can be changed slowly in time at the wave maker mimicking the amplitude modulation (e.g., Chabchoub et al., 2012, Chapter 13 of this book). The other is to control the phases of the spectral components at the wave maker such that they follow the nonlinear evolution of a known nonlinear wave field (e.g., Houtani et al., 2018, Chapter 10 of this book). It is interesting to note that the local characteristics of the extreme waves resemble each other, and either linearly focused or nonlinearly self-focused, they may result in a similar ship response (Houtani et al., 2022).

Warning criteria

From an engineering point of view, what is the most relevant information? Is the maximum individual wave height the most important parameter? Is the occurrence probability the most important parameter? Is the generation mechanism important or not? The answer depends on the purpose of each engineering field. Commercial vessels avoid storms (e.g., Sasmal et al., 2021). But fishing vessels get in trouble because they may not even receive up-to-date information about the storm. Conveying the possible location and time of the dangerous and freakish sea state is most relevant for ships. On the other hand, offshore platforms cannot avoid storms. The structure needs to sustain the most severe storms in its lifetime, say 30 to 50 years. Operations need to be stopped when a storm passes by. Knowledge of a typical strength and duration of a storm event is essential information. For that, long-term statistics need to be combined with short-term statistics (see Chapter 9 for a comprehensive review).

In the last decades, the Benjamin–Feir index (BFI) (Janssen, 2003) was studied extensively that combines representative values of waves (H_s and T_p) and parameters that characterize the spectral geometry (frequency bandwidth $\delta f/f$). Physically, the index represents the relative significance of nonlinearity and dispersion. The same parameter (inverted) appears in the theory of the stability of the Stokes wave (Benjamin and Feir, 1967). Therefore, the BFI is considered representative of the likelihood of modulational instability: $\mathrm{BFI} \equiv \sqrt{2}ak/\left(\frac{\delta f}{f}\right)$. The directional spreading of the spectrum (σ_θ) may be important as well (Waseda et al., 2009), and the BFI was extended to include directional spreading (e.g., Mori et al., 2011). Regardless of how these parameters are related to the generation mechanism of the freak wave, these parameters concisely characterize the geometry of the directional spectrum. Steepness ak is related to wave age (Toba, 1973, $ak \propto \left(c_p/u_*\right)^{-1/2}$), so the numerator of the BFI represents an inverse of a wave age (c/U_{10}); the larger the steepness, the younger the waves. The denominator of the BFI $\left(\delta f/f\right)$ more or less represents the spectral peak enhancement ($\gamma \geq 1$) and that is also related to the wave age; the younger the wave, the peak is more enhanced. Therefore, the BFI is higher for young waves. Quite often, marine accidents occur when the waves are growing, that is, when waves are young. That is partly because the ship departs when the weather is good, and accidents occur during the rapid development of the storm.

The relevance of these parameters has been debated in the past decade or so from a viewpoint of the generation mechanism (see, e.g., Chapter 3). However, from an engineering point of view, the generation mechanism may not be as relevant as one would think. We should revisit these parameters and investigate how they could be used to indicate the representation of the spectral geometry.

What is relevant from an engineering perspective?

What is relevant from an engineering perspective? The examples given earlier of the ringing of the TLP and whipping of a ship both indicate that there was an impulsive forcing on the structure. Hydrodynamic impulsive forcing likely implies that the motion of the fluid particle is accelerated, and is associated with breaking or an inception to breaking. At the stagnation point on the structure, the dynamic pressure is high, and therefore the wave imposes a localized impulsive force on the structure, just like hitting the structure with a hammer. The process of localized pressure force is the same as the impact of freak waves on breakwaters (Chapter 12). However, when it comes to coastal applications, the bathymetric effect undoubtedly becomes dominant, and the third-order nonlinearity may not be as important as in deepwater (Chapter 7).

Before breaking, the wavefront leans forward, and the front trough becomes deeper than the rear trough. However, as the wave height further increases, the geometry changes, and eventually, the rear trough becomes deeper than the front trough. When the ship encounters the wave at this moment, the ship's bow rapidly sinks into the water, and a large slamming impact force will be imposed on the ship (see Chapter 10).

These examples highlight the relevance of local geometry and kinematics from an engineering perspective. Moreover, recent studies revealed that these local features may not depend on the generation mechanism (Houtani et al., 2022). This is still an open question, but if this is indeed the case, it is also true that the generation mechanism cannot be inferred from a single-point measurement. Spatial information is necessary.

A weak nonlinear process is known as a trigger to generate a localized and highly nonlinear wave group. It is also known that the dispersive focusing will eventually create a localized wave group. The distinction, however, may exist and is related to the lifetime of the nonlinear wave group (Fig. 2.2). For offshore structures, the lifetime or the coherence timescale of a wave group may not be relevant as the wave group immediately moves away from the structure. However, for ships, if a coherent wave group can be detected by marine radar, it may be possible to avoid such a wave group. The coherency of a wave group in the ocean is not well understood yet as there is a considerable lack of information on the spatiotemporal evolution of ocean waves.

Finally, to relate long-term statistics to short-term statistics, it is important to understand how the directional spectrum forms under different meteorological forcing. Studies are showing how the wave field develops under the tropical cyclone (Mori, 2012) and extratropical cyclones (Ponce de Leon and Guedes Soares, 2015), and the distinction is related to their translation speed (Kita et al., 2018). To characterize wave fields under different meteorological conditions, the freak wave indices may be useful for prediction (Chapter 8). Furthermore, how the short-time statistics change under the global warming scenario is of high interest in shipping and offshore industries (Chapter 11). The joint probability of wave height and wave period, sometimes called a scatter diagram, may change in time, and how to incorporate that in design is a practical problem to tackle.

In the following chapters, some of the outstanding questions asked in this chapter may be answered. It is up to the reader to read the textbook from a scientific point of view or from an engineering point of view, but this book should offer a broad spectrum of information useful for both science and engineering disciplines.

References

Akhmediev, N., Soto-Crespo, J.M., Ankiewicz, A., 2009. Extreme waves that appear from nowhere: on the nature of rogue waves. Physics Letters A 373 (25), 2137–2145. Available from: https://doi.org/10.1016/j.physleta.2009.04.023.

Annenkov, S., Shrira, V., 2006. Role of Non-Resonant Interactions in the Evolution of Nonlinear Random Water Wave Fields 561, 181–207. Available from: https://doi.org/10.1017/S0022112006000632.

Barratt, D., Bingham, H.B., Adcock, T.A.A., 2020. Nonlinear evolution of a steep, focusing wave group in deep water simulated with OceanWave3D. Journal of Offshore Mechanics and Arctic Engineering 142, 021201. Available from: https://doi.org/10.1115/1.4044989.

Benetazzo, A., Barbariol, F., Bergamasco, F., Torsello, A., Carniel, S., Sclavo, M., 2015. Observation of extreme sea waves in a space-time ensemble. Journal of Physical Oceanography 45. Available from: https://doi.org/10.1175/JPO-D-15-0017.1.

Benjamin, T.B., Feir, J.E., 1967. The disintegration of wave trains on deep water Part 1. Theory. Journal of Fluid Mechanics 27 (3), 417–430. Available from: https://doi.org/10.1017/S002211206700045X.

Bitner-Gregersen, E.M., Fernandez, L., Lefèvre, J.M., Monbaliu, J., Toffoli, A., 2014. The North Sea Andrea storm and numerical simulations. Natural Hazards and Earth System Sciences 14 (6), 1407–1415. Available from: https://doi.org/10.5194/nhess-14-1407-2014.

Chabchoub, A., Hoffmann, N., Onorato, M., Slunyaev, A., Sergeeva, A., Pelinovsky, E., et al., 2012. Observation of a hierarchy of up to fifth-order rogue waves in a water tank. Physical Review E 86 (5), 056601. Available from: https://doi.org/10.1103/PhysRevE.86.056601.

Chaplin, J.R., Rainey, R.C.T., Yemm, R.W., 1997. Ringing of a vertical cylinder in waves. Journal of Fluid Mechanics. Available from: https://doi.org/10.1017/S002211209700699X.

Dias, F., Brennan, J., de León, S.P., Clancy, C., Dudley, J., 2015. Local analysis of wave fields produced from hindcasted rogue wave sea states. In: ASME 2015 34th International Conference on Ocean, Offshore and Arctic Engineering. American Society of Mechanical Engineers, V003T02A020. Available from: https://doi.org/10.1115/OMAE2015-41458.

Donelan, M.A., Curcic, M., Chen, S., Magnusson, A.K., 2012. Modeling waves and wind stress. Journal of Geophysical Research: Oceans 117 (C11). Available from: https://doi.org/10.1029/2011JC007787.

Donelan, M., Longuet-Higgins, M.S., Turner, J.S., 1972. Periodicity in whitecaps. Nature 239 (5373), 449. Available from: https://doi.org/10.1038/239449a0.

Donelan, M.A., Magnusson, A.K., 2017. The making of the Andrea wave and other rogues. Scientific Reports 7 (1), 1–7. Available from: https://doi.org/10.1038/srep44124.

Fedele, F., Brennan, J., Ponce de León, S., et al., 2016. Real world ocean rogue waves explained without the modulational instability. Scientific Reports 6, 1–11. Available from: https://doi.org/10.1038/srep27715.

Fujimoto, W., Waseda, T., Webb, A., 2019. Impact of the four-wave quasi-resonance on freak wave shapes in the ocean. Ocean Dynamics 69 (1), 101–121. Available from: https://doi.org/10.1007/s10236-018-1234-9.

Gemmrich, J., Cicon, L., 2022. Generation mechanism and prediction of an observed extreme rogue wave. Scientific Reports 12 (1), 1–10. Available from: https://doi.org/10.1038/s41598-022-05671-4.

Gramstad, O., Babanin, A., 2016. The generalized kinetic equation as a model for the nonlinear transfer in third-generation wave models. Ocean Dynamics 66 (4), 509–526. Available from: https://doi.org/10.1007/s10236-016-0940-4.

Gramstad, O., Stiassnie, M., 2013. Phase-averaged equation for water waves. Journal of Fluid Mechanics 718, 280–303. Available from: https://doi.org/10.1017/jfm.2012.609.

Hasselmann, K., 1962. On the non-linear energy transfer in a gravity-wave spectrum Part 1. General theory. Journal of Fluid Mechanics 12 (4), 481–500. Available from: https://doi.org/10.1017/S0022112062000373.

Hasselmann, K., et al., 1973. Measurements of wind-wave growth and swell decay during the Joint North Sea Wave Project (JONSWAP). Ergänzungsheft zur Deutschen Hydrographischen Zeitschrift, Reihe A 8, 12.

Houtani, H., Sawada, H., Waseda, T., 2022. Crest height amplification of modulated wave trains due to spectral broadening and phase-convergence. Fluids 7. Available from: https://doi.org/10.3390/xxxx.

Houtani, H., Waseda, T., Fujimoto, W., Kiyomatsu, K., Tanizawa, K., 2018. Generation of a spatially periodic directional wave field in a rectangular wave basin based on higher-order spectral simulation. Ocean Engineering 169, 428–441. Available from: https://doi.org/10.1016/j.oceaneng.2018.09.024.

Houtani, H., Waseda, T., Tanizawa, K., Sawada, H., 2019. Temporal variation of modulated-wave-train geometries and their influence on vertical bending moments of a container ship. Applied Ocean Research 86, 128–140. Available from: https://doi.org/10.1016/j.apor.2019.01.021.

Janssen, P.A., 2003. Nonlinear four-wave interactions and freak waves. Journal of Physical Oceanography 33 (4), 863–884. Available from: https://doi.org/10.1175/1520-0485(2003)33<863:NFIAFW>2.0.CO;2.

Kharif, C., Pelinovsky, E., 2003. Physical mechanisms of the rogue wave phenomenon. European Journal of Mechanics - B/Fluids 22, 603–634. Available from: https://doi.org/10.1016/j.euromechflu.2003.09.002.

Kita, Y., Waseda, T., Webb, A., 2018. Development of waves under explosive cyclones in the Northwestern Pacific. Ocean Dynamics 68 (10), 1403–1418. Available from: https://doi.org/10.1007/s10236-018-1195-z.

Krasitskii, V.P., 1994. On reduced equations in the Hamiltonian theory of weakly nonlinear surface waves. Journal of Fluid Mechanics 272, 1–20. Available from: https://doi.org/10.1017/S0022112094004350.

Levi, C., Welch, S., Fontaine, E., Tulin, M.P., 1998. Experiments on the ringing response of an elastic cylinder in breaking wave groups. 13th International Workshop on Water Waves and Floating Bodies .

Mori, N., Janssen, P.A., 2006. On kurtosis and occurrence probability of freak waves. Journal of Physical Oceanography 36 (7), 1471–1483. Available from: https://doi.org/10.1175/JPO2922.1.

Mori, N., Onorato, M., Janssen, P.A., 2011. On the estimation of the kurtosis in directional sea states for freak wave forecasting. Journal of Physical Oceanography 41 (8), 1484–1497. Available from: https://doi.org/10.1175/2011JPO4542.1.

Mori, N., 2012. Freak waves under typhoon conditions. Journal of Geophysical Research: Oceans 117 (C11). Available from: https://doi.org/10.1029/2011JC007788.

Onorato, M., Osborne, A.R., Serio, M., Cavaleri, L., Brandini, C., Stansberg, C.T., 2004. Observation of strongly non-Gaussian statistics for random sea surface gravity waves in wave flume experiments. Physical Review E 70 (6), 067302. Available from: https://doi.org/10.1103/PhysRevE.70.067302.

Onorato, M., Waseda, T., Toffoli, A., Cavaleri, L., Gramstad, O., Janssen, P.A.E.M., et al., 2009. Statistical properties of directional ocean waves: the role of the modulational instability in the formation of extreme events. Physical Review Letters 102 (11), 114502. Available from: https://doi.org/10.1103/PhysRevLett.102.114502.

Phillips, O.M., 1958. The equilibrium range in the spectrum of wind-generated waves. Journal of Fluid Mechanics 4 (4), 426–434. Available from: https://doi.org/10.1017/S0022112058000550.

Phillips, O.M., 1960. On the dynamics of unsteady gravity waves of finite amplitude Part 1. The elementary interactions. Journal of Fluid Mechanics 9 (2), 193–217. Available from: https://doi.org/10.1017/S0022112060001043.

Ponce de Leon, Guedes Soares, C., 2015. Hindcast of the Hercules winter storm in the North Atlantic. Natural Hazards 78, 1883–1897. Available from: https://doi.org/10.1007/s11069-015-1806-7.

Sasmal, K., Miratsu, R., Kodaira, T., Fukui, T., Zhu, T., Waseda, T., 2021. Statistical model representing storm avoidance by merchant ships in the North Atlantic Ocean. Ocean Engineering 235, 109163. Available from: https://doi.org/10.1016/j.oceaneng.2021.109163.

Solli, D.R., Ropers, C., Koonath, P., Jalali, B., 2007. Optical rogue waves. Nature 450, 1054. Available from: https://doi.org/10.1038/nature06402.

Stiassnie, M., 1984. Note on the modified nonlinear Schrödinger equation for deep water waves. Wave Motion 6 (4), 431–433. Available from: https://doi.org/10.1016/0165-2125(84)90043-X.

Tamura, H., Waseda, T., Miyazawa, Y., 2009. Freakish sea state and swell-windsea coupling: numerical study of the Suwa-Maru incident. Geophysical Research Letters 36 (1). Available from: https://doi.org/10.1029/2008GL036280.

Tanaka, M., 2001. Verification of Hasselmann's energy transfer among surface gravity waves by direct numerical simulations of primitive equations. Journal of Fluid Mechanics 444, 199–221. Available from: https://doi.org/10.1017/S0022112001005389.

Tayfun, Fedele, F., 2007. Wave-height distributions and nonlinear effects. Ocean Engineering 34, 1631–1649. Available from: https://doi.org/10.1016/j.oceaneng.2006.110.006.

Toba, Y., 1973. Local balance in the air-sea boundary processes. III. On the spectrum of wind waves. Journal of the Oceanographical Society of Japan 29 (5), 209–220. Available from: https://doi.org/10.1007/BF02108528.

Toffoli, A., Lefèvre, J., Bitner-Gregersen, E., Monbaliu, J., 2005. Towards the identification of warning criteria: analysis of a ship accident database. Applied Ocean Research 27, 281–291. Available from: https://doi.org/10.1016/j.apor.2006.03.003.

Trulsen, K., Borge, J.C.N., Gramstad, O., Aouf, L., Lefèvre, J.M., 2015. Crossing sea state and rogue wave probability during the Prestige accident. Journal of Geophysical Research: Oceans 120 (10), 7113–7136. Available from: https://doi.org/10.1002/2015JC011161.

Tulin, M.P., 1996. Breaking of ocean waves and downshifting. Waves and Nonlinear Processes in Hydrodynamics. , pp. 177–190. Springer, Dordrecht.

Waseda, T., In, K., Kiyomatsu, K., Tamura, H., Miyazawa, Y., Iyama, K., 2014. Predicting freakish sea state with an operational third-generation wave model. Natural Hazards and Earth System Sciences 14 (4), 945–957. Available from: https://doi.org/10.5194/nhess-14-945-2014.

Waseda, T., Kinoshita, T., Tamura, H., 2009. Evolution of a random directional wave and freak wave occurrence. Journal of Physical Oceanography 39 (3), 621–639. Available from: https://doi.org/10.1175/2008JPO4031.1.

Waseda, T., Tamura, H., Kinoshita, T., 2012. Freakish sea index and sea states during ship accidents. Journal of Marine Science and Technology 17 (3), 305–314. Available from: https://doi.org/10.1007/s00773-012-0171-4.

Waseda, T., Watanabe, S., Fujimoto, W., Nose, T., Kodaira, T., Chabchoub, A., 2021. Directional coherent wave group from an assimilated nonlinear wave field. Frontiers in Physics . Available from: https://doi.org/10.3389/fphy.2021.622303.

Welch, S., Levi, C., Fontaine, E., Tulin, M.P., 1999. Experimental study of the ringing response of a vertical cylinder in breaking wave groups. International Journal of Offshore and Polar Engineering 9 (04).

Yuen, H.C., Lake, B.M., 1982. Nonlinear dynamics of deep-water gravity waves. Advances in Applied Mechanics 22, 67–229. Available from: https://doi.org/10.1016/S0065-2156(08)70066-8.

Zakharov, V.E., 1967. The instability of waves in nonlinear dispersive media. Soviet Physics – Journal of Experimental and Theoretical Physics 24 (4), 740–744.

Zakharov, V.E., 1968. Stability of periodic waves of finite amplitude on the surface of a deep fluid. Journal of Applied Mechanics and Technical Physics 9 (2), 190–194. Available from: https://doi.org/10.1007/BF00913182.

Long-term in situ measurements of rogue waves

Marios Christou[1] *and Kevin Ewans*[2,3]

[1]Civil & Environmental Engineering Department, Imperial College London, London, United Kingdom [2]MetOcean Research, New Plymouth, New Zealand [3]Department of Infrastructure Engineering, University of Melbourne, Melbourne, VIC, Australia

Introduction

As discussed in Chapter 1, until late in the 19th century, the existence of rogue waves was based on observations, photographs, and mariners' tales of rogue waves from ocean-going vessels (Nikolkina and Didenkulova, 2011). Indeed, for some time the occurrence of rogue waves was thought to be a nautical myth. However, the measurement of the Draupner New Year wave in 1995 provided clear evidence that rogue waves actually exist (Haver and Anderson, 2000), but this was a one-off measurement of a single rogue wave event, and the need for corroborating evidence from more measurements was widely recognized. In fact, Haver and Anderson (2000) commented that the scarcity of rogue wave measurements makes it difficult to determine whether they conform to standard statistical models. Accordingly, large databases of field measurements that store the raw time history of the water surface elevation were needed; fortunately, in the 20 years since the Draupner New Year wave event was reported, such databases have become available, mainly due to readily available low-cost storage, enabling detailed analysis of rogue waves.

The classification of whether a large wave is a rogue wave is not clear-cut, with several definitions proposed in the literature (as discussed in Chapter 1), making comparisons of results from various measurement campaigns difficult. However, the criteria proposed by Haver (2000) are definitive and were adopted by Christou and Ewans (2014); a wave with a crest elevation η_c or height H, occurring within a 20-minute sample record, is classified as a rogue wave if

$$\frac{\eta_c}{H_s} > 1.25 \quad \text{and/or} \quad \frac{H}{H_s} > 2 \tag{3.1}$$

Science and Engineering of Freak Waves.
DOI: https://doi.org/10.1016/B978-0-323-91736-0.00001-8

where H_s is the significant wave height. In most measurement studies, the significant wave height is estimated from the zeroth moment of the wave spectrum, H_{m0}. In some studies, the average of the highest one-third of the waves in the record, $H_{1/3}$, is employed instead, but the use of $H_{1/3}$ results in less-restrictive rogue wave definitions, as $H_{1/3}$ is typically lower than H_{m0}, resulting in more rogue waves being identified. The matter is further complicated if the sample record lengths are different from 20 minutes. In addition, while most studies consider either a large crest or a large height as being sufficient to define a rogue wave, Tomita and Kawamura (2000) required that both criteria for η_c and H are satisfied to designate a rogue wave.

Field measurements are inevitably associated with instrument and data acquisition errors, requiring a comprehensive and strict quality control (QC) process to produce a reliable database that can be trusted; even if highly sophisticated analysis is performed, incorrect conclusions will be drawn from unreliable datasets. We refer the reader to Christou and Ewans (2014) and Häfner et al. (2021a,b) for a detailed description of QC procedures.

In this chapter, we deal with location-specific, long-term, in situ field measurements that have been associated with rogue waves. These types of acquisitions are often referred to as "single-point" measurements, and our focus is on field data recorded by either a moored buoy or an instrument, such as a downward-looking radar, mounted on a fixed platform. These instruments measure the time history of the water surface elevation and provide essential inputs for calibrating hindcast models and predicting extreme waves (see Chapter 9). These measurements are distinct from, but complementary to, the space-time field measurements discussed in Chapter 4. As wave measurements from buoys are based on the principle that these buoys follow the particle motion of the water surface, they are regarded as Lagrangian measurements, while measurements made from fixed instruments are regarded as Eulerian measurements; these are discussed in separate sections below. We conclude with a discussion of key findings that have resulted from such measurement programs.

Lagrangian measurements

Many routine and operational wave-measuring programs employ surface-following buoys, which are therefore an important source of long-term measurements for investigating rogue waves. Wave buoys measure the vertical acceleration, which is twice integrated to give displacement. In the case of the Datawell Waverider Buoy, which is widely regarded as the benchmark for wave height and directional measurements in the field, the vertical accelerometer on the

Waverider buoy is mounted on a gravity-stabilized platform, to compensate for the effects of roll and pitch rotations. After calibration, the buoy can measure vertical displacement values to an accuracy of 0.5% for values between ± 20 m and has a resolution of 0.01 m.

Surface-following wave buoys provide Lagrangian measurements of the sea surface, and while these result in reliable estimates of the significant wave height, wave period, and other sea-state parameters, they appear to underestimate the crest elevation statistics (e.g., Seymour and Castel, 1998; Krogstad and Barstow, 2000; Casas-Prat and Holthuijsen, 2010; Gibson et al., 2014). Some authors also postulate that in very large waves, buoys may skirt around the crest within directional sea states or be dragged through the crest itself (e.g., Forristall, 2000), but it appears that references to this effect have been conjectural, based on circumstantial evidence, and no conclusive evidence has been reported. For example, Barbariol et al. (2019) compared H_{max}/H_s values derived from a buoy that was both moored in and then freely drifting in the Southern Ocean and found the values from the free buoy tending to be $2\% - 3\%$ higher than those from the moored buoy, and concluded that the results were consistent with behavior expected from a mooring system. Such mooring constraint might also be expected to be more severe in extreme sea states and result in relatively fewer rogue wave occurrences, but this is at odds with Baschek and Imai (2011), who found that the occurrence of rogue waves was more or less independent of the significant wave height, in an analysis of 81 years of buoy data off the West Coast of the United States. This is also corroborated by Häfner et al. (2021a,b) who have more recently concluded from buoy measurements that rogue wave risk does not depend on the significant wave height.

Despite their shortcomings in measuring crest elevations, buoy data account for a large proportion of long-term datasets and can still provide valuable information on the occurrence of rogue wave events. Indeed, there are numerous reports of rogue waves being found in buoy datasets (e.g., Fedele et al., 2016; Gemmrich and Thomson, 2017; Onorato et al., 2021; Ji et al., 2022), and Table 3.1 lists a number of other long-term wave buoy datasets associated with rogue wave measurements. Fedele et al. (2016) examined three famous rogue wave events, namely, the Killard rogue wave, which was measured with a Waverider buoy offshore west Ireland, and the Draupner and Andrea rogue wave events, which were both measured with downward-looking lasers in the North Sea. They concluded that the main generation mechanism for rogue waves is constructive interference of elementary waves enhanced by second-order bound nonlinearities and not modulational instability, implying that rogue waves are likely to be rare occurrences of weakly nonlinear random seas. Gemmrich and Thomson (2017) also concluded that random superposition is sufficient to generate rogue waves, from the analysis of data that included around 4 years of Datawell Directional Waverider data (Table 3.1).

Table 3.1 Long-term Lagrangian datasets from wave buoys used to study rogue waves. The final column presents the rogue wave likelihood based on whether the crest (η_c) or height (H), or both criteria from Eq. (3.1) are met.

References	Instrument	Location	Depth (m)	Number of waves (sample duration)	H_s definition	Rogue wave likelihood
Pinho et al. (2004)	HPR buoy	Atlantic, east coast of Brazil	$1050-1250$	1.3×10^{6a} (17 min)	$H_{1/3}$	H: 8.5×10^{-5}
Baschek and Imai (2011)	Datawell Directional Waverider buoy Mk-II, Mk-III	Six shallow water locations—Pacific, West Coast, United States	$16-40$	1.8×10^{8} (30 min)	H_{m0}	H: 1.4×10^{-5}
		Five coastal ocean locations—Pacific, West Coast, United States	$113-363$	2.2×10^{8} (30 min)	H_{m0}	H: 7.7×10^{-6}
		Five open ocean locations—Pacific, West Coast, United States	$319-550$	2×10^{8} (30 min)	H_{m0}	H: 1.4×10^{-5}
Candella (2016)	Triaxys buoys	Three locations—Atlantic, east coast of Brazil	200	4.9×10^{6} (20 min)	$H_{1/3}$	H: 1.4×10^{-4}
Gemmrich and Thomson (2017)	Datawell Waverider buoy	One location—Northeast Pacific	Not given	6.1×10^{6} (40 min)	H_{m0}	H: 5.4×10^{-6b} η_c: 1.1×10^{-5}
Cattrell et al. (2018)	Datawell Waverider buoys	80 locations—North America and Pacific Ocean island coasts	$10-4500$	$\times10^{9}$ (20 min)	H_{m0}	H: 6.8×10^{-5} η_c: 2.0×10^{-5}
Häfner et al. (2021a, b)	Datawell Waverider buoys Mk-III (mostly), Mk-IV	158 locations—North America and Pacific Ocean island coasts	Not given	1.9×10^{9} (30 min)	H_{m0}	H: 6.6×10^{-5}

[a] *Assuming 6 s mean zero-crossing period per sea-state sample.*
[b] *Using a definition of $H/H_s > 2.2$*

One of the earliest observations of the occurrence of rogue waves in such a database, following the reporting of the Draupner New Year wave event, was reported by Pinho et al. (2004). They found 276 rogue wave occurrences in 7457 datasets, occurring in both severe storms and moderately calm ocean surface conditions. They also estimated kurtosis values of sample records associated with rogue waves to be between 3 and 3.6 and estimated JONSWAP peak enhancement factors to be between 1 and 3, but these values are also common for sample records without rogue waves (e.g., Christou and Ewans, 2014; Amurol et al., 2014).

Baschek and Imai (2011) observed 7157 rogue waves in a particularly large database of buoy measurements off the West Coast of the United States, made in "shallow" water, "coastal" ocean, and "open" ocean locations (Table 3.1), consisting of 80.8 years of data, following an extensive QC. They found that the occurrence of rogue waves was more or less independent of the significant wave height, but overall occurrences were lower than reported by others.

Candella (2016) identified more than 700 rogue waves in 12,620, 20-minute Triaxys buoy sample records off the east coast of Brazil, using the wave height criterion in Eq. (3.1) with $H_s = H_{1/3}$. Neither the vertical asymmetry (the ratio of the crest to trough) nor the directional spreading (derived from the spectra) was found to have a significant correlation with the occurrence of rogue waves. Candella (2016) did not provide details of their QC, except to mention that they eliminated records with spurious data and sea states in which $H_{1/3}$ was less than 0.5 m.

Cattrell et al. (2018) analyzed a very large dataset of surface wave following buoys (Table 3.1), with the objective of identifying predictors of rogue wave occurrence. They found trends in the occurrence of rogue waves, which followed a power law relationship with mean significant wave height and mean zero up-crossing wave period, and concluded that the predictors are dependent on geographical location. The dataset was subjected to a strict QC process, based on that reported by Christou and Ewans (2014).

Cattrell et al. (2019) analyzed data recorded with Datawell Waverider buoys off the West Coast of the United States, spanning 23 years, and found that rogue wave occurrence generally reduced as a function of time while the severity (or relative height) was rising. They also found a seasonal dependence, with winter generating a greater occurrence and higher severity of rogue waves, which they hypothesize is potentially due to increased spectral bandwidth. The same QC procedure, as outlined in Cattrell et al. (2018), was applied to the data.

Häfner et al. (2021a,b) also had the objective of identifying rogue wave indicators, in their analysis of a very large buoy dataset (Table 3.1). Using robust Bayesian statistics and machine learning, they found relevant sea-state parameters for rogue wave heights and crests. In the case of wave heights, the

Benjamin–Feir index, steepness, and kurtosis were found to be weak predictors, whereas crest–trough correlation was the dominating rogue wave predictor regardless of the location, water depth, and environmental conditions. In the case of crests, they find that the crest–trough and spectral bandwidth had very low predictive power, whereas surface elevation skewness, steepness, and Ursell number had the highest predictive power. However, it is important to note that they mentioned the same shortcomings of crest measurements from wave buoys as discussed above. Overall, they conclude that rogue waves are generated by linear superposition in bandwidth-limited seas and that the correction due to nonlinearities is small. The Free Ocean Wave Dataset used for the study involved a QC process based on Christou and Ewans (2014).

It is clear from the foregoing that wave buoys provide useful data for rogue wave investigations. The wave height criterion in Eq. (3.1) can be used unambiguously to detect rogue wave events in these datasets, but waves meeting the crest criterion in Eq. (3.1) have also been detected (Cattrell et al., 2018, 2019). Accordingly, several conclusions have been drawn, as noted above, but a common conclusion, especially resulting from the studies based on a substantial number of buoy measurements, is that constructive interference of waves from the random wave field is the main generating mechanism, as concluded by Christou and Ewans (2014).

Eulerian measurements

There have been several studies employing instruments mounted on fixed structures to measure the water surface elevation offshore, including wave radars, lasers, and staffs, among others. These are referred to as Eulerian measurements, and in their review article, Dysthe et al. (2008) stated that rogue wave measurements from fixed instruments are the most reliable. In recent years, the wave radar has seen increased adoption across the world, with the Rosemount WaveRadar REX (previously Saab WaveRadar REX, and hereafter referred to as REX radar) in particular being adopted on the majority of fixed offshore platforms (Christou and Ewans, 2014). There are many studies that have investigated a single or few rogues waves, such as Walker et al. (2004); Cherneva and Guedes Soares (2008); Adcock et al. (2011); Fedele et al. (2016); and Donelan and Magnusson (2017), among others. However, this chapter is focused on long-duration datasets of rogue waves measured from fixed instruments, and Table 3.2 presents a summary of these studies.

The REX radar employs a frequency-modulated continuous microwave signal that linearly sweeps between 9.7 and 10.3 GHz. The frequency of the signal reflected from the water surface is different from the transmitted signal, which evolves over time due to the linear sweep. The distance between the REX radar and the water surface elevation (referred to as the *range*) can therefore be

Table 3.2 Long-term Eulerian datasets from fixed instruments used to study rogue waves. The final column presents the rogue wave likelihood based on whether the crest (η_c), height (H), or both criteria from Eq. (3.1) are met.

References	Instrument	Location	Depth (m)	Number of waves (sample duration)	H_s definition	Rogue wave likelihood
Olagnon and van Iseghem (2000)	Plessey wave radar	North Sea	106	1.6×10^6 (20 min)	$H_{1/3}$	H: 5.062×10^{-5} η_c: 4.758×10^{-5}
Stansell (2004)	Thorn EMI laser	North Sea	130	0.35×10^6 (20 min)	$H_{1/3}$	H: 2.197×10^{-4}
Krogstad et al. (2008)	Optech Laser Array	North Sea	70	2.9×10^{6}[a,c] (20 min)	H_{m0}	H: 3.6×10^{-5} η_c: 2.9×10^{-5} H and η_c: 1.7×10^{-5}
Christou and Ewans (2014)	Rosemount WaveRadar REX (principally[b])	North Sea (principally[b])	$7.7 - 1311$	122×10^6 (20 min)	H_{m0}	H or η_c: 2.991×10^{-5}
Makri et al. (2016)	Rosemount WaveRadar REX	North Sea	$85 - 186$	184×10^6 (20 min)	H_{m0}	H or η_c: 5.4×10^{-5}
Kvingedal et al. (2018)	Rosemount WaveRadar REX	North Sea	190	42×10^6 (20 min)	H_{m0} (assumed)	H: 4.0×10^{-5} η_c: 1.6×10^{-5}
Malila et al. (2023)	Optech Laser Array	North Sea	70	7×10^{6}[c] (20 min)	H_{m0}	H: 4.0×10^{-5} η_c: 8.5×10^{-5}
Teutsch and Weisse (2023)	Rosemount WaveRadar REX	North Sea	$21 - 34$	36.8×10^{6}[a] (30 min)	$H_{1/3}$	H: 1.3×10^{-4} η_c: 2.6×10^{-5} H and η_c: 3.9×10^{-5}

[a] *Assuming 6 s mean zero-crossing period per sea-state sample.*
[b] *Christou and Ewans (2014) also examined a small number of sea states from the following instruments: EH radar, Marex radar, Optech laser, EMI laser, wave staff, and step gauge; they also investigated a small number of sea states from the Gulf of Mexico, South China Sea, and North West Shelf of Australia.*
[c] *Only considered sea states with $H_s > 3$ m.*

determined, as it is proportional to the difference between the transmitted and reflected frequencies. The REX radar is calibrated before installation in a special facility, and its accuracy is ± 6 mm (range <50 m) or ± 12 mm (range >50 m); importantly for operations, it does not need recalibration. It is a common misconception that the REX radar takes the average water surface elevation over a footprint generated by a ± 5 degree beamwidth. As explained by Ewans et al. (2014), in reality the process is more complicated and dependent on the intensity and angle of the transmitted microwave signal, the direction of the reflected signal, signal attenuation, and processing methodology. Jangir et al. (2022) investigated the wave radar performance using numerical simulations and determined that the actual REX radar (that they referred to as the "radar-dominant signal") performs significantly better than taking the average water level over the footprint, which underestimates the surface elevation with a mean absolute bias of approximately 0.3 m. Consequently, the REX radar performs better than might be expected based on a simple interpretation of measurement footprint. The operation of other wave radars is very similar to that of the REX, and the reader is referred to Christou and Ewans (2014) for a more detailed description of the various instruments used for Eulerian measurements.

Christou and Ewans (2014) performed a comprehensive study of rogue waves measured principally by the REX radar from fixed jacket structures. They collated a large dataset of 122 million waves with significant wave heights ranging from 0.12 to 15.4 m and peak periods from 1 to 24.7 s. As most of the dataset was recorded in the southern and central North Sea, their findings represent waves propagating in intermediate water depths within extratropical locations. The dataset was divided into 20-minute sea-state samples, and after a strict, no-tolerance QC process was applied, there were 528,475 *normal* wave samples and 3649 *rogue* wave samples, where the latter were distinguished using Eq. (3.1) with $H_s = H_{m0}$. The aim of the study by Christou and Ewans (2014) was to identify mechanisms that can generate rogue waves in the field, and consequently, the sea-state parameters, environmental conditions, and local individual wave parameters were examined, the results of which will be summarized within the following paragraphs.

Regarding the sea-state parameters, Fig. 3.1A presents the significant wave height against the peak period for every normal wave sample (gray dots) against those for the rogue wave samples (black dots) (Christou and Ewans, 2014). The black dots representing the rogue wave samples lie within the cloud of gray dots for the normal wave samples, suggesting that the sea-state steepness by itself cannot explain the formation of rogue waves. This result is consolidated in Fig. 3.2A, which plots the maximum crest elevation $\eta_{\max}$ against the mean sea-state steepness $S_1 = 2\pi H_s / g T_1^2$, which benefits from the increased stability of the mean wave period T_1 that is derived from the ratio of the zeroth- and first-order spectral moments. Once again, the rogue wave samples are located within the cloud representing normal wave samples. Fig. 3.2B considers the skewness and

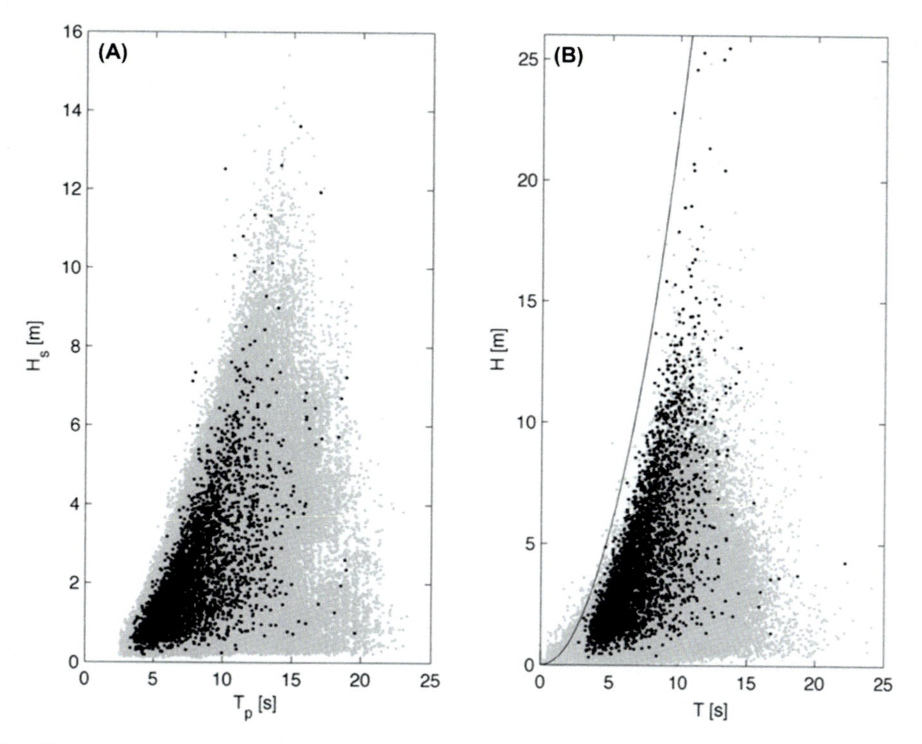

Figure 3.1 (A) Significant wave height versus peak period indicating sea-state steepness for 20-min normal wave samples (gray dots) and rogue wave samples (black dots). (B) Individual wave height versus period indicating individual wave steepness for normal waves (gray dots) and rogue waves (black dots) and 1/7th steepness curve (black line). *Source: Reproduced from Christou, M., Ewans, K., 2014. Field measurements of rogue water waves. Journal of Physical Oceanography 44(9), 2317–2335. https://doi.org/10.1175/JPO-D-13-0199.1. © American Meteorological Society. Used with permission.*

illustrates that sea states with the same skewness can lead to both rogue wave and normal wave samples, and therefore skewness is not a predictor of rogue waves. Fig. 3.3 examines whether kurtosis can be a predictor for rogue waves. Panel (A) shows that rogue wave samples appear to have kurtosis values greater than three, which is generally larger than those of the normal wave samples. However, this is misleading, as a rogue wave sample will by definition lead to a large kurtosis because the rogue wave is significantly larger than the background waves. Therefore, Stansell (2004) proposed first removing the rogue wave event from the time trace and then computing the kurtosis of the remaining sea state. Christou and Ewans (2014) applied this same procedure, which results in Fig. 3.3B and demonstrates that rogue wave samples have the same range of kurtosis values as experienced by normal wave samples; consequently, kurtosis is not a rogue wave indicator, which is in agreement with Stansell (2004) and more recently Fedele et al. (2016) and Häfner et al. (2021a,b).

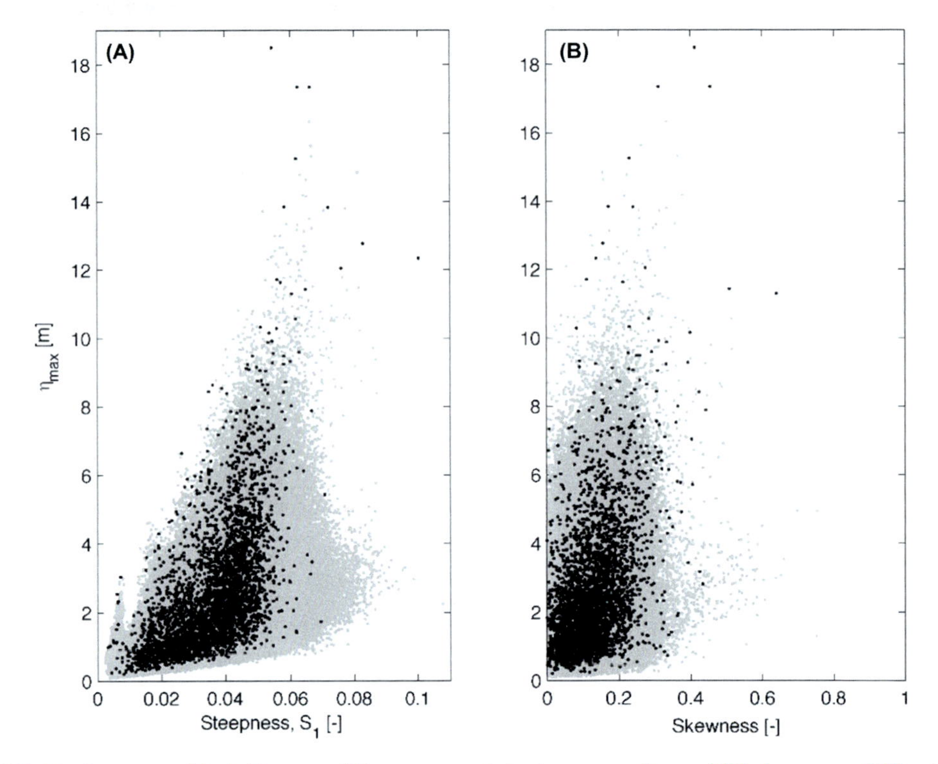

Figure 3.2 Maximum crest height versus (A) mean sea-state steepness, S_1, and (B) skewness of 20-min normal wave samples (gray dots) and rogue wave samples (black dots). *Source: Reproduced from Christou, M., Ewans, K., 2014. Field measurements of rogue water waves. Journal of Physical Oceanography 44(9), 2317–2335. https://doi.org/10.1175/JPO-D-13-0199.1. © American Meteorological Society. Used with permission.*

Christou and Ewans (2014) went on to investigate whether the environmental conditions (winds, waves, and currents) in which rogue waves appeared could explain their generation. The wind and wave fields were extracted from an Oceanweather Inc. hindcast (Cardone and Cox, 2011), and the frequency-directional wave spectra were partitioned into wind wave and swell modes using XWaves (Hanson and Phillips, 2001). The tidal currents were calculated using POLPRED (Proctor et al., 2004), and the residual current was assumed to have a negligible influence, as the measurements were recorded in the North Sea. For data recorded at the Goldeneye platform (central North Sea in a water depth of 122 m with 77,083 normal wave samples and 743 rogue wave samples), Fig. 3.4 illustrates the empirical probability density functions of the absolute environmental conditions, and Fig. 3.5 presents the equivalent for the relative conditions. As can be observed in both these figures, the distributions for the rogue wave samples appear remarkably similar to those for the normal waves; note that the noisy rogue wave distributions are a result of two orders

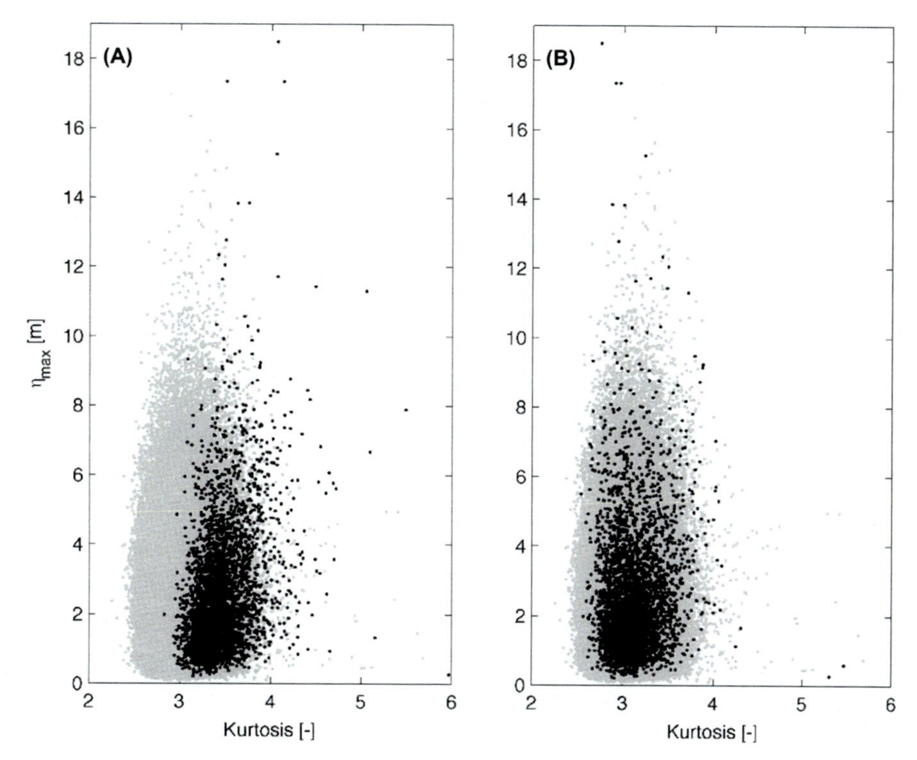

Figure 3.3 Maximum crest height versus kurtosis of 20-min wave samples. Subfigure (A) illustrates the normal wave samples (gray dots) and rogue wave samples with the rogue wave event present (black dots). Subfigure (B) presents the normal wave samples (gray dots) and rogue wave samples with the individual rogue wave event removed from the 20-min sample (black dots); the kurtosis is then calculated, and as such it is not biased by the presence of a rogue wave. *Source: Reproduced from Christou, M., Ewans, K., 2014. Field measurements of rogue water waves. Journal of Physical Oceanography 44(9), 2317–2335. https://doi.org/10.1175/JPO-D-13-0199.1. © American Meteorological Society. Used with permission.*

of magnitude less data than the normal wave samples. Consequently, the environmental conditions do not seem to contribute to the rogue wave generation.

Christou and Ewans (2014) then investigated the influence of local, individual wave parameters on rogue waves. Fig. 3.1B presents the height H against the period T for individual waves, and it illustrates that while rogue waves tend to be steep, there are at least as many normal waves that are equally steep. Therefore the local wave steepness cannot be the sole driver of rogue waves. This agrees with more recent work by Fedele et al. (2016) and Häfner et al. (2021a,b), who have concluded that rogue waves are enhanced or corrected by second-order nonlinearities, in contrast to being dominated by modulational instabilities. Christou and Ewans (2014) then considered the normalized average shape of the 3649 rogue waves and whether it differs from the largest 1% of normal waves (corresponding to 5286 waves). Fig. 3.6 demonstrates that

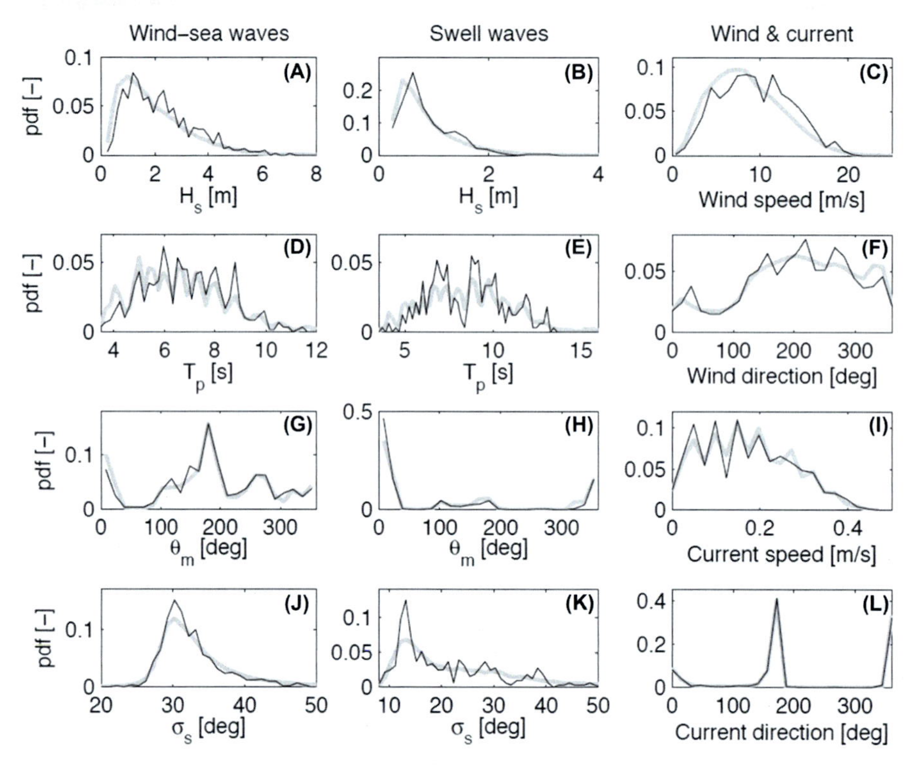

Figure 3.4 Empirical probability density functions of the absolute environmental conditions for normal wave samples (gray line) and rogue wave samples (black line). The significant wave height (H_s), peak period (T_p), mean direction of propagation (θ_m), and directional spreading (σ_s) for the wind sea components are presented in the first column (panels (A), (D), (G), and (J)) and for the swell partition in the second column (panels (B), (E), (H), and (K)). The third column (panels (C), (F), (I), and (L)) illustrates the wind speed and direction as well as the current speed and direction. *Source: Reproduced from Christou, M., Ewans, K., 2014. Field measurements of rogue water waves. Journal of Physical Oceanography 44(9), 2317–2335. https://doi.org/10.1175/JPO-D-13-0199.1. © American Meteorological Society. Used with permission.*

rogue waves also follow the autocorrelation NewWave shape (Lindgren, 1970; Boccotti, 1983, 2000; Tromans et al., 1991), which is yet another similarity between rogue and normal waves. However, the rogue waves exhibit higher crests and deeper troughs when compared to the largest normal waves. This means that nonlinearity alone cannot be the difference between rogue waves and normal waves, as in that case the crests would be higher but the troughs shallower; indeed, as discussed above, Fig. 3.1B demonstrates that normal waves are just as steep as rogue waves. Importantly, the NewWave shape also indicates that rogue waves are generated by constructive interference caused by dispersive focusing. Therefore, the larger crests and deeper troughs exhibited by the rogue waves may simply indicate a greater degree of constructive interference with more frequency components coming into phase with each other.

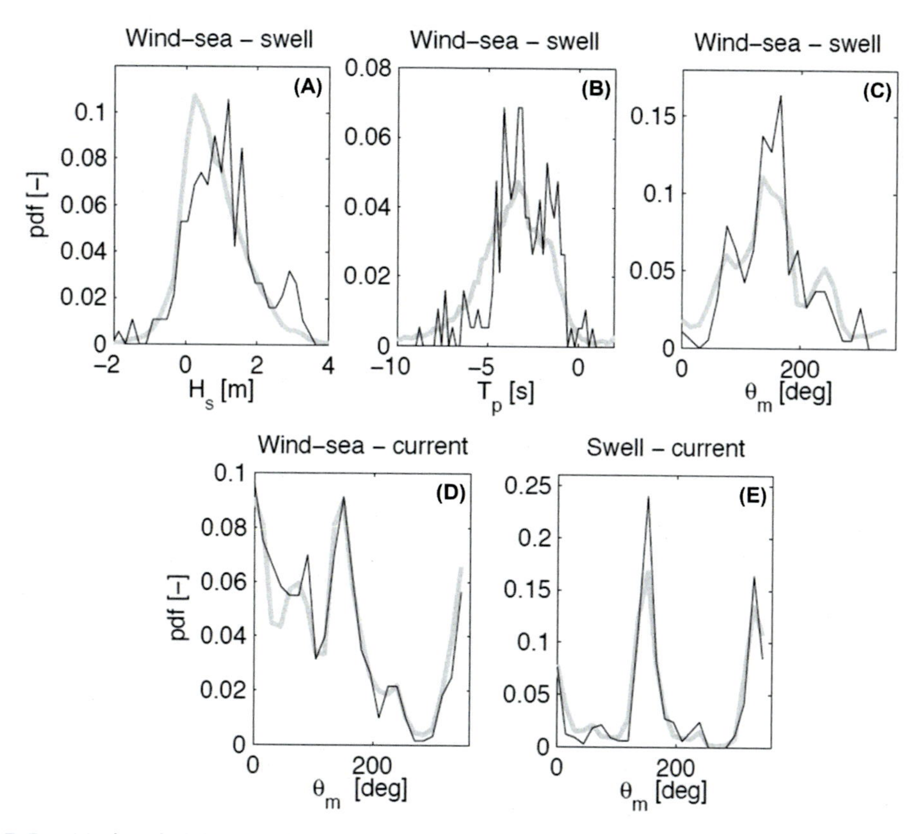

Figure 3.5 Empirical probability density functions of the relative environmental conditions for normal wave samples (gray line) and rogue wave samples (black line). (A) Significant wave height (H_s) from wind sea minus swell partitions; (B) peak period (T_p) from wind sea minus swell partitions; (C) mean direction of propagation (θ_m) from wind sea minus swell partitions; (D) mean direction from wind sea minus current; and (E) mean direction from swell minus current. *Source: Reproduced from Christou, M., Ewans, K., 2014. Field measurements of rogue water waves. Journal of Physical Oceanography 44(9), 2317–2335. https://doi.org/10.1175/JPO-D-13-0199.1. © American Meteorological Society. Used with permission.*

To examine dispersive focusing further, Christou and Ewans (2014) used a wavelet transform (with Morlet wavelet) to investigate the behavior of rogue waves in the local vicinity of their occurrence. Fig. 3.7 presents the variance density spectrum S and phase ϕ of three representative rogue waves at the time of the event as well as one wave period before and after the rogue wave. These panels illustrate that at the time of the rogue wave event (Fig. 3.7D–F), the majority of the frequencies are in phase with each other, leading to constructive interference and thus a large wave. The constructive interference may not always lead to a focused crest event with $\phi \approx 0$ degrees, as in Fig. 3.7D, but the frequencies will coalesce around the same phase at the time of the rogue wave, such as $\phi \approx 25$ degrees in panel (E) or $\phi \approx 75$ degrees in panel (F). Examining the

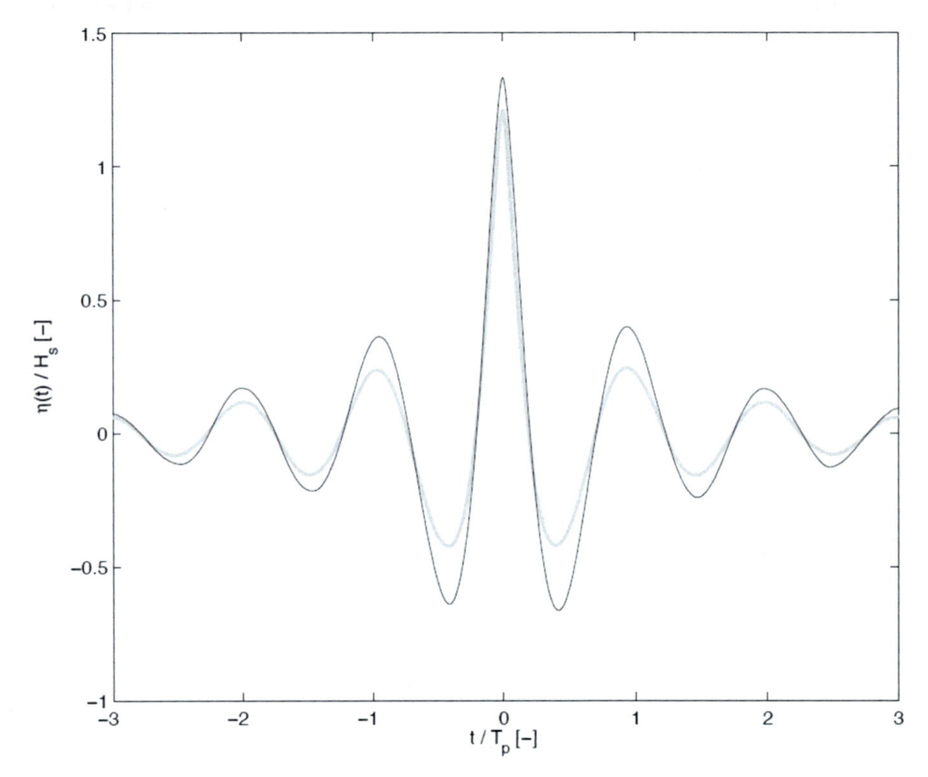

Figure 3.6 Average wave shape of the highest 1% of crests in normal sea states (gray line) and from all rogue crests (black line). *Source: Reproduced from Christou, M., Ewans, K., 2014. Field measurements of rogue water waves. Journal of Physical Oceanography 44(9), 2317–2335. https://doi.org/10.1175/JPO-D-13-0199.1. © American Meteorological Society. Used with permission.*

phases before and after the rogue wave event demonstrates the familiar pattern from the linear dispersion process within focused wave events; this leads to a positive linear gradient in the phases before focusing (Fig. 3.7A–C) and a negative linear gradient after focusing (Fig. 3.7G–I).

Christou and Ewans (2014) subsequently compared the degree of constructive interference of all 3649 rogue waves against the largest 1% of normal waves. This was achieved by calculating the standard deviation of the phases σ_{phase} for a range of frequency components: a small value of σ_{phase} indicates a wave event that is undergoing constructive interference, as the variation in phase values is low, and therefore most frequencies have approximately the same phase. The number of frequency components considered in the calculation of σ_{phase} was varied and only included frequencies whose variance density was within a certain α percentage of the spectral peak S_p. For example, a value of $\alpha = 5\%$ corresponds to the j frequencies with variance density $S(f_j) > 0.95S_p$, which only

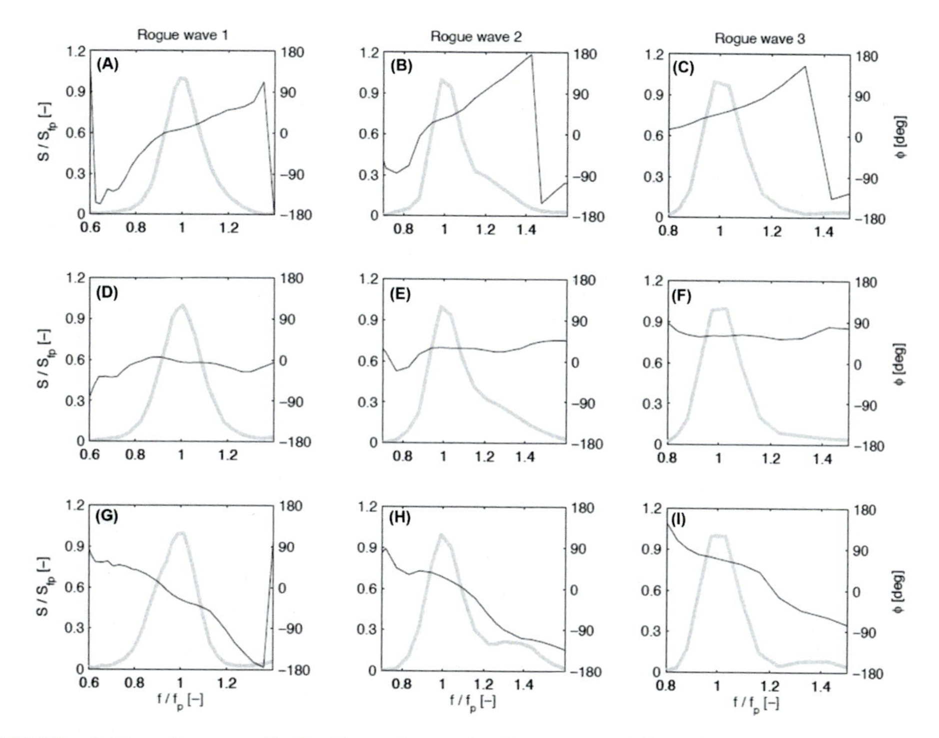

Figure 3.7 Wavelet transform results for the variance density spectra, S (gray line), and phases, ϕ (black line), of three rogue wave events. The left axis indicates the variance density spectrum normalized with respect to the peak value, and the right axis represents the phase in degrees. (A), (B), and (C) occur for one wave period before the rogue wave event; (D), (E), and (F) are at the time of the rogue wave event; and (G), (H), and (I) are for one wave period after the rogue wave event. *Source: Reproduced from Christou, M., Ewans, K., 2014. Field measurements of rogue water waves. Journal of Physical Oceanography 44(9), 2317–2335. https://doi.org/10.1175/JPO-D-13-0199.1. © American Meteorological Society. Used with permission.*

represents the most energetic frequency components, most likely located around the spectral peak. As α is gradually increased, this considers more and more of the spectral frequencies and hence is a better test of constructive interference. Fig. 3.8 illustrates the empirical probability distribution of σ_{phase} for a variety of α values for the rogue waves and largest 1% of normal waves. This figure demonstrates that for small values of α the distributions for rogue waves and the largest normal waves are very similar, which is as expected given that this is only considering a small number of frequencies close to the spectral peak. In contrast, as the value of α increases, it is evident that the rogue waves distribution has smaller σ_{phase} values than that of the largest normal waves, which indicates that the rogue waves are experiencing a higher degree of constructive interference, even out to the high frequencies. While the differences may seem small, it is important to note that these differences are sufficiently large for the rogue waves to exceed the criteria stated in Eq. (3.1).

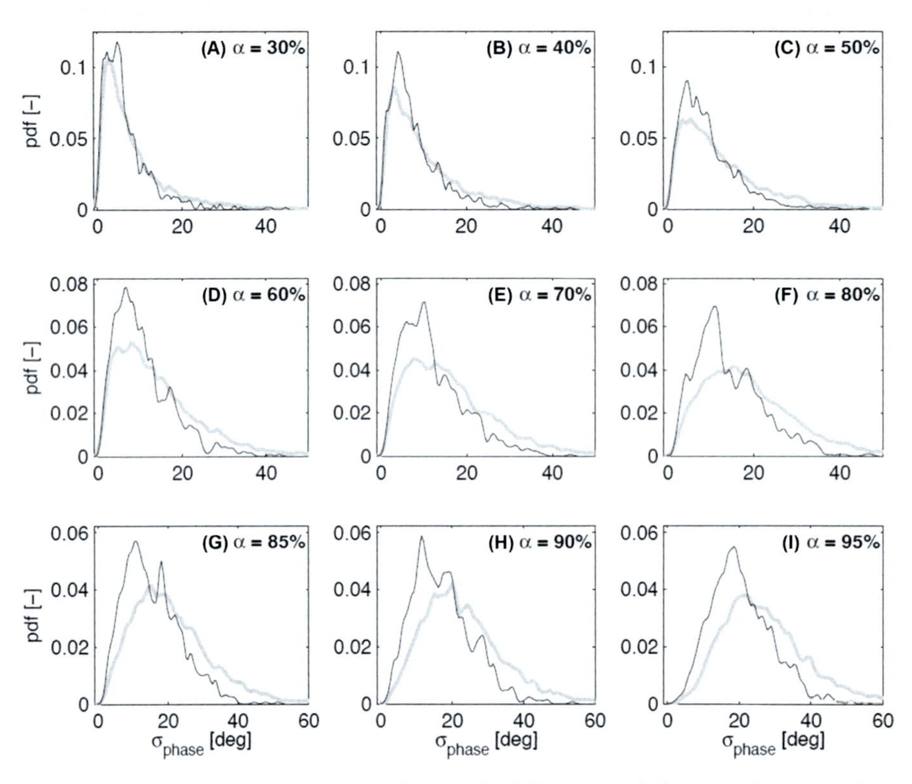

Figure 3.8 Empirical probability distributions of the standard deviation of phases at the time of a maximum event, σ_{phase}, for the highest 1% of normal waves (gray line) and rogue wave events (black line). The various sub-figures present the distributions for different values of α, as given in the top right of each of the panels (A) to (I). *Source: Reproduced from Christou, M., Ewans, K., 2014. Field measurements of rogue water waves. Journal of Physical Oceanography 44(9), 2317–2335. https://doi.org/10.1175/JPO-D-13-0199.1. © American Meteorological Society. Used with permission.*

In summary, Christou and Ewans (2014) provided evidence to support the notion that rogue waves are simply rare occurrences of the normal population, which are generated by constructive interference caused by dispersive focusing. More recent studies, such as Fedele et al. (2016); Gemmrich and Thomson (2017); Häfner et al. (2021a,b); and Teutsch and Weisse (2023), have also concurred with this key finding.

Discussion

This chapter has considered several long-term, single-point field datasets recorded using floating wave buoys (Lagrangian measurements) and fixed instruments such as wave radars (Eulerian measurements). With this wealth of information, it is worthwhile comparing the likelihood of observing a rogue wave based on whether the crest (η_c), height (H), or both criteria from Eq. (3.1) are met.

Table 3.1 presents a summary of the long-term Lagrangian datasets used to study rogue waves, and Table 3.2 presents the equivalent for the long-term Eulerian datasets. The final columns of these tables present the likelihood of a rogue wave occurring, based on the whole dataset examined by each study. It is important to note that these tables only pertain to long-duration datasets and that there are additional studies not listed (e.g., Casas-Prat and Holthuijsen, 2010; Cattrell et al., 2019, among others), as only those studies from which the rogue wave probabilities could be definitively ascertained were included.

Examining the probability of observing a rogue wave height from Table 3.1, it is clear that the range in values between the studies is notable, with a difference of two orders of magnitude. Clearly, these probabilities are a function of the number of rogue waves identified within each study, which is dependent on the duration of the sea-state sample and how the normalizing significant wave height H_s in Eq. (3.1) is calculated, both of which are also listed in the tables. However, even after being corrected, somewhat crudely, to be consistent with Eq. (3.1) using $H_s = H_{m0}$ (by multiplying the likelihood by the ratio of a 20-minute record length to the actual record length, by 0.95 if $H_s = H_{1/3}$, and by 1.1 if the wave height criterion of $H > 2.2H_s$ has been used) the vast difference in values persists. However, this range is substantially reduced to one order of magnitude, if only the very large datasets of Baschek and Imai (2011); Cattrell et al. (2018); and Häfner et al. (2021a,b) are considered, which have a mean value of 2.7×10^{-5}. Examining Table 3.2 from the Eulerian measurements, it is interesting to note that the range of probabilities of observing a rogue are much tighter, with approximately a factor of two difference (when excluding the much shorter Stansell, 2004, dataset) rather than several orders of magnitude. The similarity in the rogue wave likelihoods in Table 3.2 may be explained by the fact that all the datasets were recorded in the North Sea, whereas Table 3.1 considers field measurements from a range of locations. Finally, considering the mean value of 2.7×10^{-5} discussed above from the longest Lagrangian datasets, it is remarkably similar to the value of 2.991×10^{-5} found by Christou and Ewans (2014), which is one of the longest Eulerian datasets.

In summary, these long-term measurements have provided invaluable evidence regarding the generation mechanism and probability of observing rogue waves in the field. With the benefit of a long dataset, several of these studies have dismissed the traditional conclusions based on a single (often Draupner) or a handful of rogue wave rogue events. In particular, there appears to be no dependence of rogue occurrence on the traditional sea-state parameters nor on the kurtosis of the surface elevation, which is merely high due to the presence of the rogue wave itself and not due to the background sea state (Stansell, 2004; Christou and Ewans, 2014; among others). In recent years, several measurement campaigns provide evidence to support the generation mechanism of constructive interference from dispersive focusing. This fundamental linear process of wave dispersion is then amplified by second-order (or weak) nonlinearities leading to wave amplifications beyond the typical Gaussian approximation (Christou and Ewans, 2014; Fedele et al., 2016; Gemmrich and Thomson, 2017; Häfner et al., 2021a,b; Teutsch and Weisse, 2023). However, it must be noted that Cattrell et al. (2018) demonstrated that rogue waves also depend on the geographic location, and there are other location-specific generation mechanisms that must be considered, such as modulational instability (see Chapter 5), crossing seas (see Chapter 6), and wave–current interactions and bathymetric focusing (see Chapter 7).

References

Adcock, T.A.A., Taylor, P.H., Yan, S., Ma, Q.W., Janssen, P.A.E.M., 2011. Did the Draupner wave occur in a crossing sea? Proceedings of the Royal Society A467, 3004–3021. Available from: https://doi.org/10.1098/rspa.2011.0049.

Amurol, S., Ewans, K., Sheikh, R., 2014. Measured wave spectra offshore Sabah & Sarawak, Offshore Technology Conference, Kuala Lumpur, Malaysia, 25–28 March 2014.

Barbariol, F., Benetazzo, A., Bertotti, L., Cavaleri, L., Durrant, T., McComb, P., et al., 2019. Large waves and drifting buoys in the Southern Ocean. Ocean Engineering 172, 817–828.

Baschek, B., Imai, J., 2011. Rogue wave observations off the US West Coast. Oceanography 24 (2), 158–165. Available from: https://doi.org/10.5670/oceanog.2011.35.

Boccotti, P., 1983. Some new results on statistical properties of wind waves. Applied Ocean Research 5, 134–140. Available from: https://doi.org/10.1016/0141-1187(83)90067-6.

Boccotti, P., 2000. Wave Mechanics for Ocean Engineering. Elsevier, 496 pp.

Candella, R.N., 2016. Rogue waves off the south/southeastern Brazilian coast. Natural Hazard 83, 211–232. Available from: https://doi.org/10.1007/s11069-016-2312-2.

Cardone, V.J., Cox, A.T., 2011. Modelling very extreme sea states (VESS) in real and synthetic design level storms, Proceedings, 30th International Conference on Offshore Mechanics and Arctic Engineering, vol. 2. American Society of Mechanical Engineers, Rotterdam, Netherlands, pp. 531–544.

Casas-Prat, M., Holthuijsen, L.H., 2010. Short-term statistics of waves observed in deep water. Journal of Geophysical Research 115, C09024. Available from: https://doi.org/10.1029/2009JC005742.

Cattrell, A.D., Srokosz, M., Moat, B.I., Marsh, R., 2018. Can rogue waves be predicted using characteristic wave parameters? Journal of Geophysical Research: Oceans 123, 5624–5636. Available from: https://doi.org/10.1029/2018JC013958.

Cattrell, A.D., Srokosz, M., Moat, B.I., Marsh, R., 2019. Seasonal intensification and trends of rogue wave events on the US western seaboard. Scientific Reports 9, 4461. Available from: https://doi.org/10.1038/s41598-019-41099-z.

Cherneva, Z., Guedes Soares, C., 2008. Non-linearity and non- stationarity of the New Year abnormal wave. Applied Ocean Research 30, 215–220. Available from: https://doi.org/10.1016/j.apor.2008.08.003.

Christou, M., Ewans, K., 2014. Field measurements of rogue water waves. Journal of Physical Oceanography 44 (9), 2317–2335. Available from: https://doi.org/10.1175/JPO-D-13-0199.1.

Donelan, M., Magnusson, A.K., 2017. The making of the Andrea wave and other rogues. Scientific Reports 7, 44124. Available from: https://doi.org/10.1038/srep44124.

Dysthe, K., Krogstad, H.E., Müller, P., 2008. Oceanic rogue waves. Annual Review of Fluid Mechanics 40, 287–310. Available from: https://doi.org/10.1146/annurev.fluid.40.111406.102203.

Ewans, K., Feld, G., Jonathan, P., 2014. On wave radar measurement. Ocean Dynamics 64, 1281–1303.

Fedele, F., Brennan, J., Ponce de Leon, S., Dudley, J., Dias, F., 2016. Real world ocean rogue waves explained without the modulation instability. Scientific Reports. Available from: https://doi.org/10.1038/srep27715.

Forristall, G.Z., 2000. Wave crest distributions: observations and second-order theory. Journal of Physical Oceanography 30, 1931–1943. Available from: https://doi.org/10.1175/1520-0485(2000)030,1931:WCDOAS.2.0.CO;2.

Gemmrich, J., Thomson, J., 2017. Observations of the shape and group dynamics of rogue waves. Geophysical Research Letters 44, 1823–1830. Available from: https://doi.org/10.1002/2016GL072398.

Gibson, R., Christou, M., Feld, G., 2014. The statistics of wave height and crest elevation during the December 2012 storm in the North Sea. Ocean Dynamics 64, 1305–1317.

Häfner, D., Gemmrich, J., Jochum, M., 2021a. Real-world rogue wave probabilities. Scientific Reports 11, 10084. Available from: https://doi.org/10.1038/s41598-021-89359-1.

Häfner, D., Gemmrich, J., Jochum, M., 2021b. FOWD: a free ocean wave dataset for data mining and machine learning. Journal of Atmospheric and Oceanic Technology 38, 1305–1322. Available from: https://doi.org/10.1175/JTECH-D-20-0185.1.

Hanson, J., Phillips, O., 2001. Automatic analysis of ocean surface directional wave spectra. Journal of Atmospheric and Oceanic Technology 18, 277–293. Available from: https://doi.org/10.1175/1520-0426(2001)018,0277.

Haver, S., 2000. Evidences of the existence of freak waves. Proceedings of the Rogue Waves. Ifremer, Brest, France, pp. 129–140.

Haver, S., Anderson, O.J., 2000. Freak waves: rare realization of a typical population or typical realizations of a rare population? Proceedings of the 10th International Offshore and Polar Engineering Conference. International Society of Offshore and Polar Engineers, Seattle, WA, pp. 123–130.

Jangir, P.K., Ewans, K.C., Young, I.R., 2022. On the functionality of radar and laser ocean wave sensors. Journal of Marine Science and Engineering 10, 1260. Available from: https://doi.org/10.3390/jmse10091260.

Ji, X., Li, A., Li, J., Wang, L., Wang, D., 2022. Research on the statistical characteristic of freak waves based on observed wave data. Ocean Engineering 243, 110323.

Krogstad, H.E., Barstow, S.F., 2000. A unified approach to extreme value analysis of ocean waves. Proceedings of the 10th International Offshore and Polar Engineering Conference. International Society of Offshore and Polar Engineers, Seattle, WA, pp. 103–108.

Krogstad, H.E., Barstow, S.F., Mathiesen, L.P., Lønseth, L., Magnusson, A.K., Donelan, M.A., 2008. Extreme waves in the long-term wave measurements at Ekofisk. Proceedings of the Rogue Waves 2008 Workshop. Ifremer, Brest, France, pp. 23–33.

Kvingedal, B., Bruserud, K., Nygaard, E., 2018. Individual wave height and wave crest distributions based on field measurements from the northern North Sea. Ocean Dynamics 68 (12), 1727–1738.

Lindgren, G., 1970. Some properties of a normal process near a local maximum. Annals of Mathematical Statistics 41, 1870–1883. Available from: https://doi.org/10.1214/aoms/1177696688.

Makri, I.M., Rose, S.M., Christou, M., Gibson, R., Feld, G., 2016. Examining field measurements of deep-water crest statistics. Proceedings of the 35th Conference of Ocean, Offshore and Arctic Engineering, Busan, South Korea.

Malila, M.P., Barbariol, F., Benetazzo, A., Breivik, Ø., Magnusson, A.K., Thomson, J., et al., 2023. Statistical and dynamical characteristics of extreme wave crests assessed with field measurements from the North Sea. Journal of Physical Oceanography 53 (2), 509–531.

Nikolkina, I., Didenkulova, I., 2011. Rogue waves in 2006–2010. Natural Hazards and Earth System Sciences 11, 2913–2924. Available from: https://doi.org/10.5194/nhess-11-2913-2011.

Olagnon, M., van Iseghem, S., 2000. Some observed characteristics of sea states with extreme waves. Proceedings of the 10th International Offshore and Polar Engineering Conference. International Society of Offshore and Polar Engineers, Seattle, WA, pp. 84–90.

Onorato, M., Cavaleri, L., Randoux, S., Suret, P., Ruiz, M., de Alfonso, M., et al., 2021. Observation of a giant nonlinear wave-packet on the surface of the ocean. Scientific Reports. Available from: https://doi.org/10.1038/s41598-021-02875-y.

Pinho, U.F., Liu, P.C., Ribeira, C.E.P., 2004. Freak waves at Campos basin, Brazil. Geofizika 21, 53–66.

Proctor, R., Bell, C., Eastwood, L., Holt, J.T., Prandle, D., Young, E.F., 2004. UK marine renewable energy atlas: phase 2—POL contribution. Proudman Oceanographic Laboratory Internal Rep. 163, 26 pp.

Seymour, R.J., Castel, D., 1998. Systematic underestimation of maximum crest heights in deep water using surface-following buoys. Proceedings of the 17th International Conference on Offshore Mechanics and Arctic Engineering. ASME, Lisbon, Portugal, pp. 1–8.

Stansell, P., 2004. Distributions of freak wave heights measured in the North Sea. Applied Ocean Research 26, 35–48. Available from: https://doi.org/10.1016/j.apor.2004.01.004.

Teutsch, I., Weisse, R., 2023. Rogue waves in the Southern North Sea—the role of modulational instability. Journal of Physical Oceanography 53 (1), 269–286.

Tomita, H., Kawamura, T., 2000. Statistical analysis and inference from the in-situ data of the Sea of Japan with relevance to abnormal and/or freak waves. Proceedings of the 10th International Offshore and Polar Engineering Conference. International Society of Offshore and Polar Engineers, Seattle, WA, pp. 116–122.

Tromans, P.S., Anaturk, A., Hagemeijer, P., 1991. A new model for the kinematics of large ocean waves—application as a design wave. Proceedings of the First International Offshore and Polar Engineering Conference. International Society of Offshore and Polar Engineers, Edinburgh, Scotland, pp. 64–71.

Walker, D.A.G., Taylor, P.H., Eatock-Taylor, R., 2004. The shape of large surface waves on the open sea and the Draupner New Year wave. Applied Ocean Research 26, 73–83. Available from: https://doi.org/10.1016/j.apor.2005.02.001.

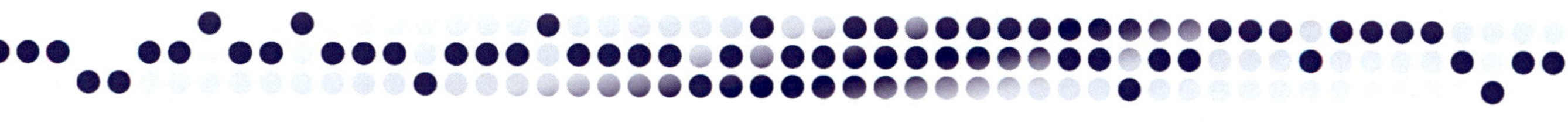

Measurements 2: space-time measurements of freak waves

Alvise Benetazzo[1], *Filippo Bergamasco*[1,2] *and Francesco Barbariol*[1]

[1]Institute of Marine Sciences (ISMAR), National Research Council (CNR), Venice, Italy

[2]University of Venice "Ca' Foscari,", Venice, Italy

Introduction

Measurement of freak waves is challenging, especially in the open ocean. There is a vast amount of wave records containing freak waves during marine storms, made mainly by devices (e.g., buoys or wave probes) providing the 1-D profile of the wave elevation η as a function of time t at a single point $P = (x_p, y_p)$ of the sea surface (x and y are two axes of a horizontal coordinate plane). However, oceanic waves that can be observed on the surface of oceans have 3-D (2-D + time) shape $\eta(x, y, t)$ that changes randomly (Figs. 4.1 and 4.2), and there is not an *a priori* method that determines the "where" and "when" a freak wave event may occur. Indeed, individual wind-generated waves have short lengths and crests (scales of tens to hundreds of meters) and are short-living (tens of seconds), and the physical processes involved range from mechanics of wave motion to forcing processes (such as wind and currents). Consequently, only a single freak wave is usually reported in most registered time-series events, proving the strong localization of the wave energy of the largest waves. Therefore there is an evident need for modern instrumentations and tools capable of accurately recording large waves not only in time but also in space and detecting freak waves if they occasionally appear. This chapter outlines what concepts are required to measure 3-D freak waves, describes sensors that retrieve space-time wave fields, and shows examples of freak waves during marine storms.

Science and Engineering of Freak Waves.
DOI: https://doi.org/10.1016/B978-0-323-91736-0.00005-5

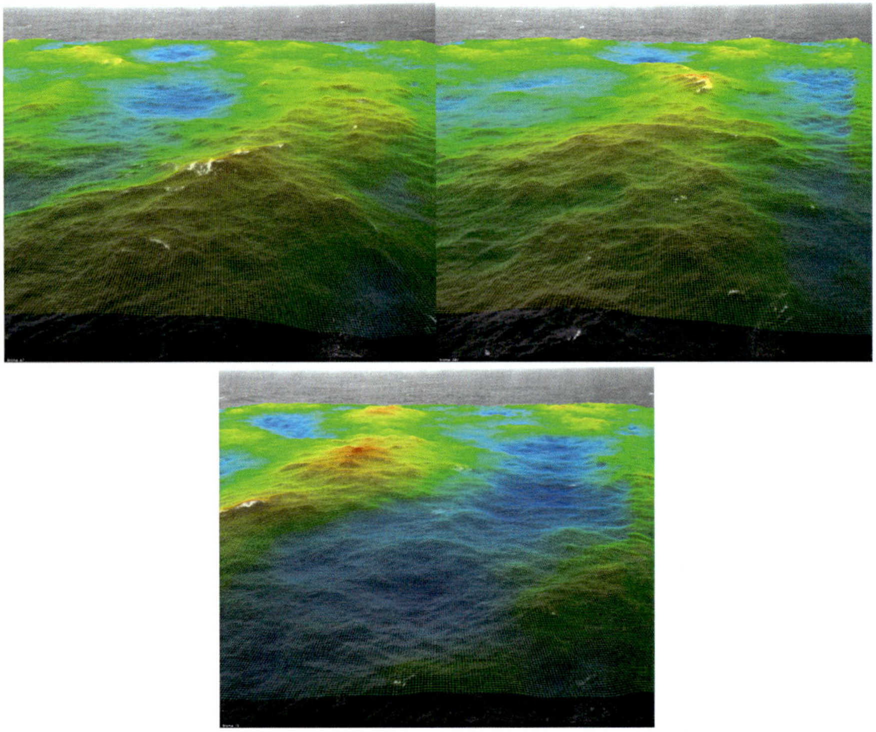

Figure 4.1 Measured wave elevation field $\eta(x, y)$ at different times t during the same sea state. The color shading is proportional to the sea elevation (the orange shows wave crests, and the blue shows wave troughs), and it overlays the gray-scale image of the sea surface. Waves propagate from left to right of the elevation maps.

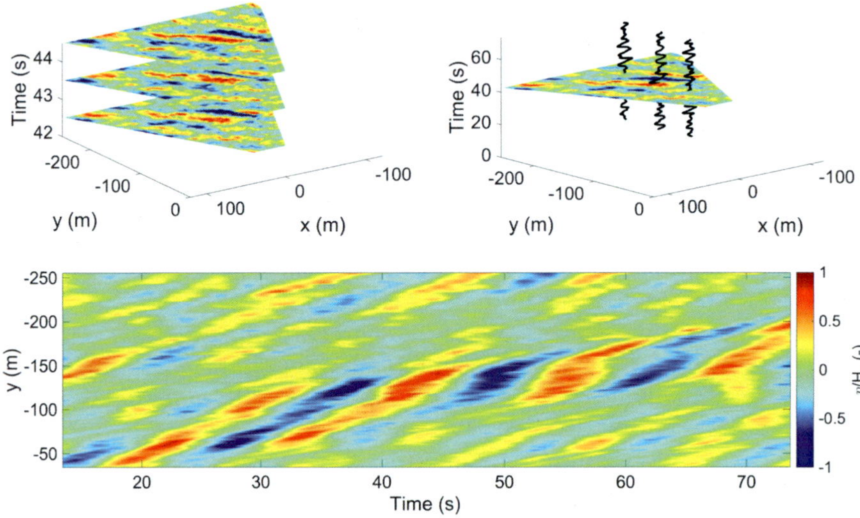

Figure 4.2 Different visualizations of the space-time wave field $\eta(x, y, t)$, normalized with the significant wave height H_s. (*top-left*) Slices of $\eta(x, y, t)$ on the horizontal *xy*-plane at different times. The color is proportional to the wave elevation. (*top-right*) Time series $\eta(t)$ at different positions $P = (x_p, y_p)$ on the horizontal *xy*-plane. (*bottom*) *yt*-diagram along the transect at $x = 0$.

Description of freak waves in space and time

The definition of whether or not a wave is a freak in space-time follows the standard criteria valid for time series, based on scaling characteristic lengths with the significant wave height H_s, i.e., wave elevation $>1.25H_s$ and/or crest-to-trough wave height $>2H_s$, where H_s is estimated from the standard deviation of $\eta(x, y, t)$ (see Chapter 1).

Early scientific observations of freak waves in 2-D space and time were made by Benetazzo et al. (2015) using stereo photography. For a sea state that is statistically homogeneous over a 2-D surface region R (for instance, the colored region in Fig. 4.1, which spans an area A of about 100×100 m^2) and stationary during a temporal interval Γ (duration of a few to tens of minutes), the procedure defines the crest height of freak waves as a local maximum η_{max} (Fig. 4.3) of the evolving field $\eta(x, y, t)$ exceeding the threshold $1.25H_s$ within the finite 2-D + time region R Γ. Analogously, freak troughs may be described as local minima over R and Γ, although there is still no classification of freak waves based on the amplitude of minima of $\eta(x, y, t)$.

Approaching the freak wave observation based on space-time measurements, the question can be answered on how localized these events are (Fig. 4.4). As a matter of fact, a single-point measurement system that is displaced from the position of the freak wave (for instance, the region bounded by the white contour in Fig. 4.3) can overlook the event (compare solid blue and red lines in Fig. 4.5). Remarkably, freak wave analysis based on single-point series underestimates the maximum amplitudes and probability that freaks can attain during a sea state since this type of record is blind in telling at what stage of the wave focusing the measurement was taken (Fedele et al., 2013).

Interpreting the amplitude and the likelihood of freak waves collected in space and time requires dedicated statistical methods. The approaches developed for multidimensional random fields applied to 3-D waves provide consistent results (Benetazzo et al., 2015; Fedele, 2012). An example is shown in Fig. 4.6. The probability distribution that best fits experimental data is based on the space-time extreme statistics and accounts for second-order nonlinear interactions; this supports the finding that large and freak waves during marine storms are generated by focusing on elementary waves enhanced by bound nonlinearities (Fedele et al., 2016).

The number of freak wave events can be remarkably high when a space-time measure is used. Benetazzo et al. (2015) observed 23 freak waves in a 30-min space-time record spanning an area A of about 3000 m^2. This result can be interpreted in the context of the "areal effect" (Forristall, 2006; Fedele et al., 2017; Malila et al., 2023), for which extreme waves taken over a 2-D region with area $A > 0$ are

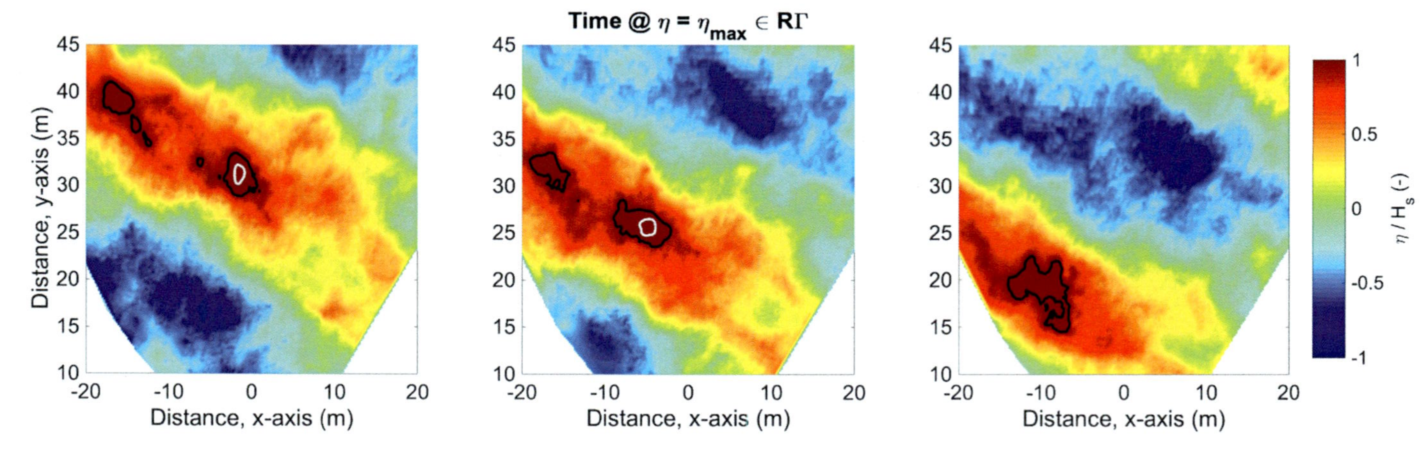

Figure 4.3 Example of freak wave detection within the space-time elevation field $\eta(x, y, t)$. Wave motion is from the top-right to the bottom-left of the elevation maps. (*middle*) A portion of the wave field $\eta(x, y)$ around the focusing point where $\eta = \eta_{\max}$ over $R\Gamma$; (*left*) growth phase a few seconds before focusing; (*right*) decay phase a few seconds after focusing. Black and white contours show the elevation levels $\eta = H_s$ and $\eta = 1.25H_s$, respectively.

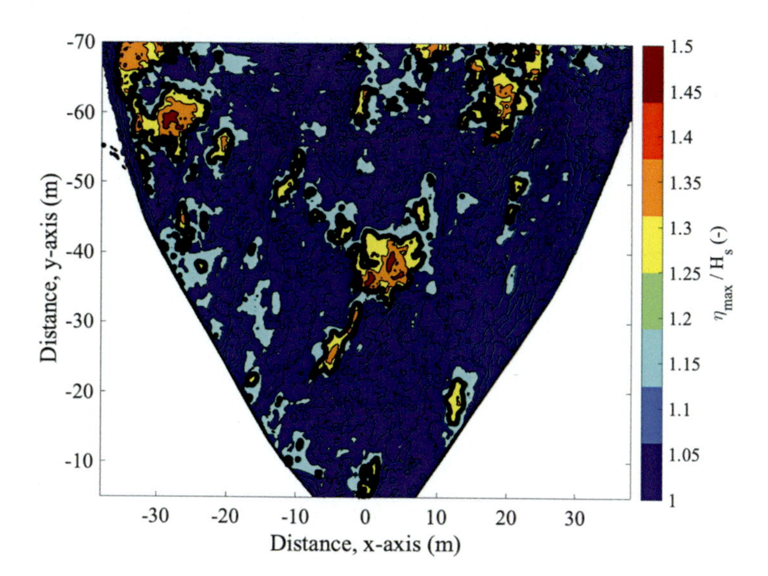

Figure 4.4 Example of the 2-D spatial position of freak waves ($\eta_{max} > 1.25H_s$; solid black contour) for a 30-min record.

larger than those at a fixed point P inside that region. Forristall (2006) used this concept to explain the damages observed on the lower decks of offshore platforms after storms. For explanation, a statistical similarity and a universal law for space-time extremes in wind seas have been suggested by Fedele et al. (2017).

For a given sea state, identifying a freak wave statistical sample can help determine the shape of its distribution. Contrary to the use of individual freak wave observations, the characterization of a distribution function is essential to determine the sea-state conditions and environmental factors favorable to the generation of extreme waves (Benetazzo et al., 2017). The sample of freaks also provides essential information on freak wave lifetime, spatial dimension, and temporal profile (Fig. 4.7).

Finally, since the definition of individual waves in space-time is uncertain, the search for freak waves based on the criterion of the maximum crest-to-trough wave height (H_{max} > *threshold*) is made through the zero-crossing analysis of all time series $\eta(t)$ with $t \in \Gamma$ over the region R. As a result, position $P_{freak} = (x_{freak}, y_{freak})$ and time t_{freak} where freak waves based on wave elevation η or wave height H criteria are detected might not coincide, conveying the evolution of space-time wave groups (Boccotti, 2000).

Sensors for space-time measurements

Measuring freak and non-freak waves in their spatial and temporal dimensions is more complicated than producing elevation time series at a single point at

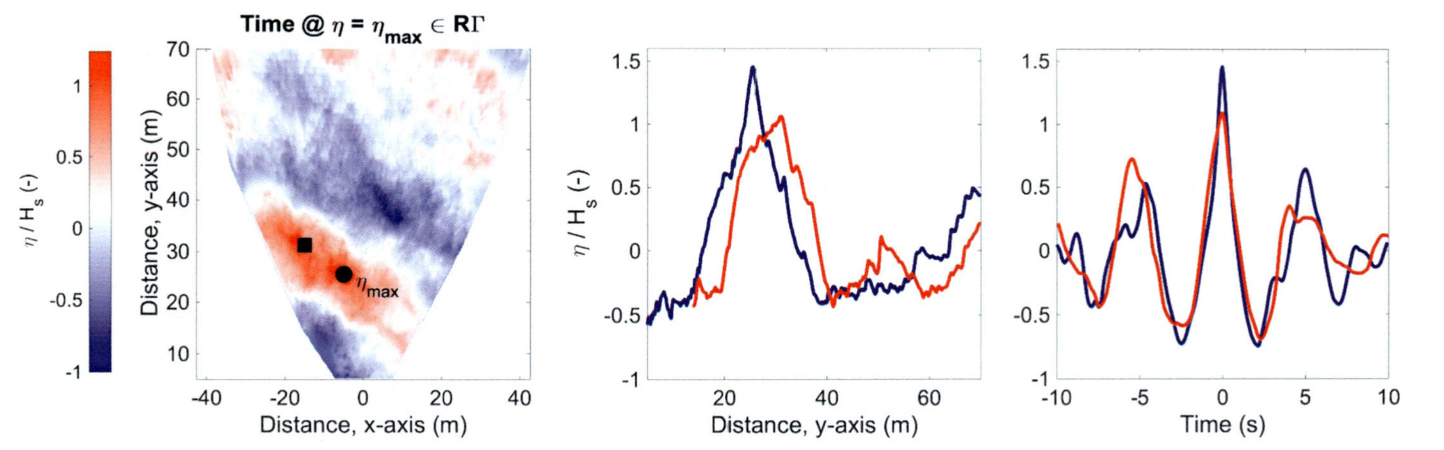

Figure 4.5 Freak wave detection in the 2-D wave elevation field $\eta(x, y)$ and wave profiles (spatial resolution: 0.2 m; temporal resolution: 0.067 s). (*left*) Wave field $\eta(x, y)$ at the time of a freak wave ($\eta_{max} > 1.25H_s$). The position of η_{max} is shown with a circle marker; the square marker shows a displaced position at a 20 m distance along the wave crest. Spatial (*middle*) and temporal (*right*) 1-D transects through the (x, y) position of η_{max} (blue line) and the displaced position (red line).

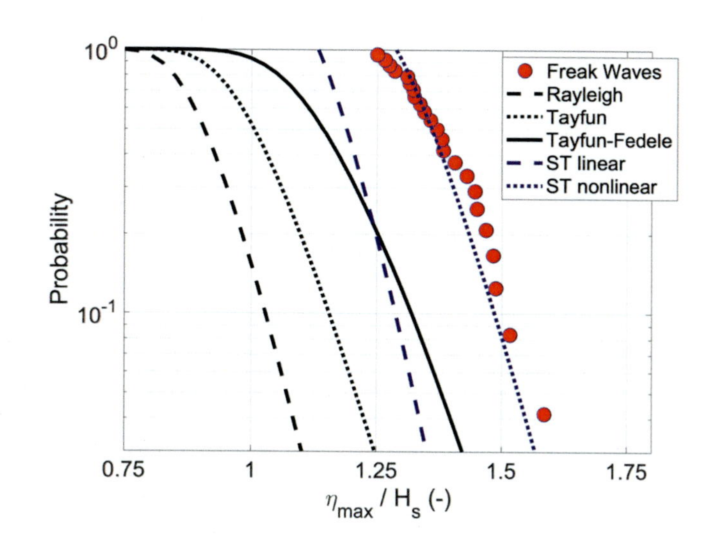

Figure 4.6 Short-term probability of space-time freak wave elevation η_{max}. Reference distributions of extremes (sea-state duration $\Gamma = 30$ min): Rayleigh, Tayfun (Tayfun, 1980), Tayfun–Fedele (Tayfun & Fedele, 2007), linear space-time (ST; Fedele, 2012), and second-order nonlinear space-time (Benetazzo et al., 2015).

sea. In general, tradeoffs have to be made depending on the desired spatial and temporal accuracy and resolution, resulting in different technologies that can be used alone or combined to exploit the attributes of each sensor.

Almost all the techniques used nowadays fall within the so-called remote sensing processes, where the physical characteristics of an area (the wave elevation in this case) are measured by capturing the reflected and/or emitted radiation at a distance (Khorram et al., 2012). Depending on the distance of the sensor (airborne, oceanographic platform, research vessel, etc.) and the kind of radiation sensed (light intensity, light polarization, X-band radio backscatter, etc.), the width of the observed area can extend from a few centimeters to kilometers. In what follows, we summarize the three major technologies involved, focusing in particular on stereo imaging which has proven in last years to be particularly effective for the space-time measurement and characterization of freak waves in the open ocean.

Stereo imaging

Driven by the advances in computer vision and 3-D reconstruction techniques, stereo imaging has gradually become a popular technique for the direct acquisition of space-time wave fields in areas of $O(100 \times 100\ \text{m}^2)$, with a spatial resolution of a point every 20–50 cm and a temporal rate up to 15 Hz. Typical accuracy for wave elevation is within 3–10 cm at close and large ranges from the cameras, respectively. Such values refer to typical installations with cameras

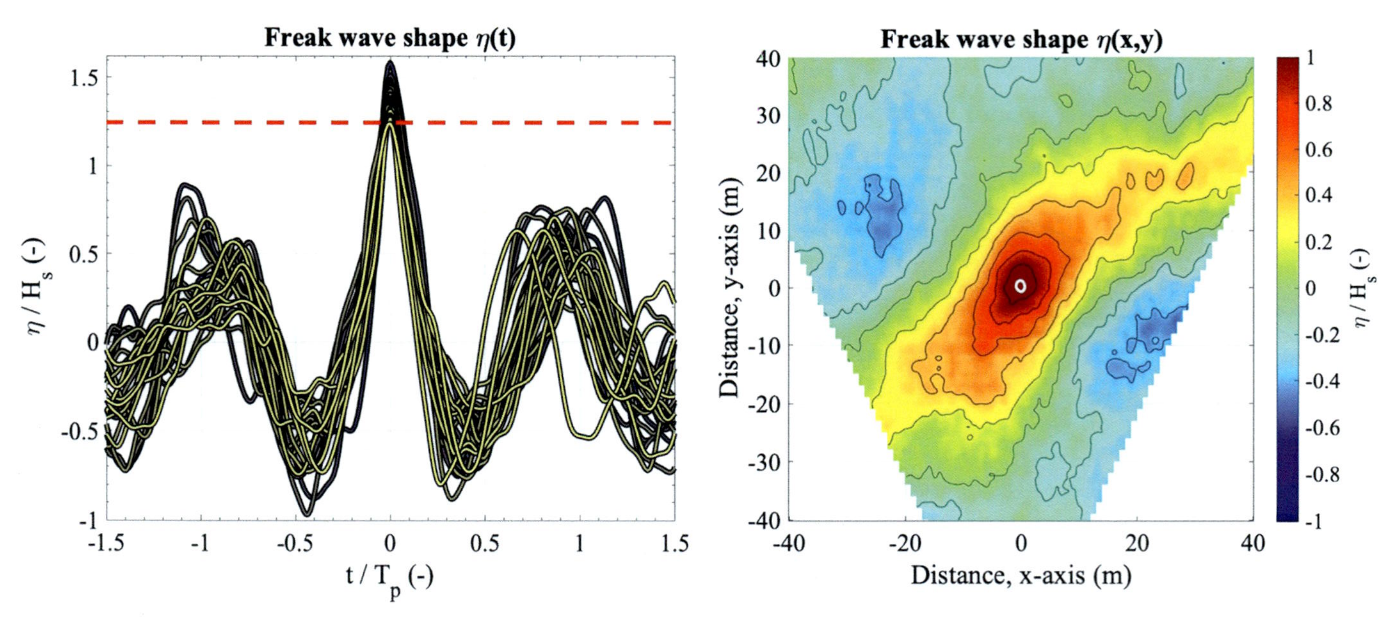

Figure 4.7 (*left*) Example of temporal profiles $\eta(t)$ of space-time freak waves ($\eta_{max} > 1.25H_s$; the dashed red line shows the threshold level) taken at different positions $P_{freak} \in R$ during the same sea state ($\Gamma = 30$ min; T_p is the peak period). The time of η_{max} is set to zero for all profiles. (*right*) Example of the average shape of the elevation field $\eta(x, y)$ at the time of the freak waves. The white contour shows the isoline $\eta = 1.25H_s$. Positions of η_{max} are set at $(x, y) = (0, 0)$, and surface elevation isolines (solid black line) are shown at 0.2 intervals.

placed 20 m above sea level but may be customized at each setup, trading off accuracy, spatial resolution, and area extent.

The hardware involved is simple as it comprises two compact industrial monochrome cameras (Fig. 4.8, left). In a typical installation, left and right imaging devices are protected from rain and sea spray and firmly mounted on a metal bar (sometimes the handrail of the oceanographic platform works just fine) and oriented in the same direction. The whole rig is tilted downward with an angle ranging from 30 to 45 degrees to the ground plane, depending on the height of the whole structure (Fig. 4.8, right).

Stereo imaging works by geometrically relating photometric features observed on images captured simultaneously from two digital cameras placed side-by-side (Fig. 4.9) at a fixed distance called baseline (2–5 m, depending on the camera-to-sea distance). Once similar points have been automatically detected in the two (stereo) frames, a process known as triangulation allows the reconstruction of the 3-D position of such points with respect to the reference frame of one of the two cameras.

For a correct reconstruction, the two cameras must capture images at the same time. Empirically, the time difference should be less than 0.1 ms to have a negligible effect on the result. For that reason, a dedicated wire connects the two devices with a trigger unit providing electrical signals to synchronize the acquisitions. Image data are transferred to a computer via fiber optics or Gigabit Ethernet and stored uncompressed to a mass storage device for further processing.

Figure 4.8
(*left*) Industrial camera (5 megapixel, 8-bit monochrome, global shutter) and lens (distortionless with 5-mm focal length) used for stereo wave imaging. (*right*) Installation of two cameras in their protective housings facing the sea surface from a research vessel.

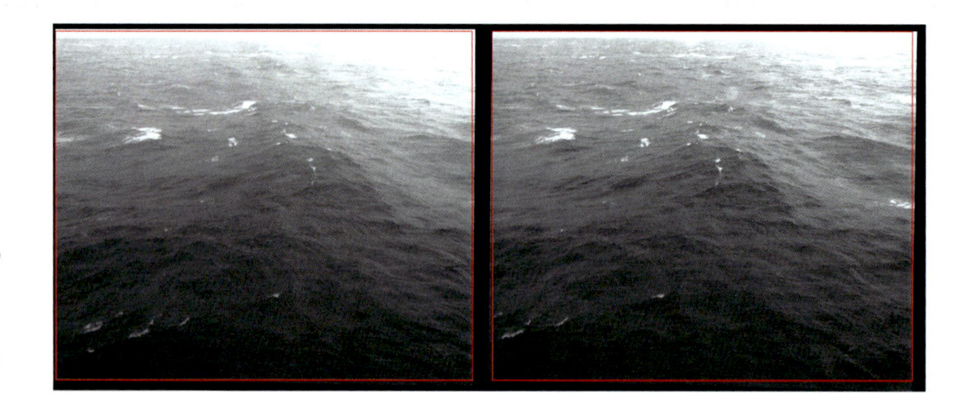

Figure 4.9 Example of stereo pair of the ocean surface grabbed by two displaced and synchronized digital cameras.

The reconstruction process is composed of a chain of image processing algorithms executed sequentially for each stereo image. For this reason, the software implementing these functions is generally referred to as a 3-D reconstruction pipeline. Among the different approaches that can be found in the literature, the open-source Wave Acquisition Stereo System (WASS; Bergamasco et al., 2017, and https://sites.google.com/unive.it/wass/home) is the most popular code. It is composed of four distinct programs to automatically analyze each stereo pair and produce a 3-D point cloud $\eta(x, y)$. The first, called *wass_prepare*, undistorts the input frame removing radial distortion introduced by the camera lenses. This step is crucial to exploit the epipolar geometry in the reconstruction stage. The second program, *wass_match*, automatically recovers the reciprocal position of the two cameras based on the image content. This information is needed for the final triangulation step. The third, called *wass_autocalibrate*, merges the reciprocal position of the two cameras among several frames to improve reliability. This operation is technically called extrinsic camera calibration and should be performed at every new acquisition. Finally, *wass_stereo* performs the so-called dense stereo reconstruction algorithm to produce a 3-D point for every image pixel. The typical output from megapixel cameras is a point cloud $\eta(x, y)$ composed of millions of 3-D points for each stereo frame. Using a temporal sequence of stereo pairs processed with the methods mentioned above provides the desired wave field $\eta(x, y, t)$.

One of the difficulties of stereo reconstruction is that the output point cloud has no uniform density in the measurement area. Indeed, for the projective nature of the geometry involved, regions close to the camera are denser than the ones farther away. Moreover, the regular image pixel grid back-projects to a nonuniform scattered cloud in the mean sea plane. Therefore, data resampling (interpolation on a regular *xy*-grid) is a mandatory step before any subsequent operation, especially if performed in the wavenumber–frequency domain. Such resampling can be computationally expensive and introduce aliasing depending on the interpolation algorithm involved. Finally, low-pass filtering is operated in the temporal domain after gridding to remove unwanted spikes from the data.

Overall, using a stereo system requires some expertise, but the resulting fields are extremely valuable to the characterization of freak waves, especially in describing their spatial characteristics. The main disadvantage of stereo imaging is that cameras can only operate during daylight since they require natural reflection from the water surface. Sun glints should be avoided since they cannot be matched. Rain and sea spray can also disturb the optical path of the reflected light from the ocean to the cameras.

Polarimetric imaging

At a centimeter scale, polarimetric imaging allows the acquisition of the surface elevation with unmatched accuracy. It works by capturing the light polarization state in the observed area to relate the so-called angle of linear polarization and degree of linear polarization with the surface slope at each point. The acquisition is performed with a single polarimetric camera with an image resolution ranging typically from 1 to 5 megapixels paired with long focal length optics. It can be mounted on oceanographic platforms or research vessels at a few meters above sea level so that the resulting captured area is about 1 m^2 (Zappa et al., 2008; Pezzaniti et al., 2009). With this technique, the elevation is not sensed directly but has to be computed with numerical integration from the slope field (Scherr, 2017). This implies that (1) the absolute elevation of each point remains unknown since the integral of a function is defined up to an unknown additive constant; (2) small biases in the slope field tend to accumulate, producing a low-frequency distortion of the resulting elevation field. This effect is particularly significant for large areas, and indeed, most of the works presented in the literature restrict the acquisition to a very small portion of the sea surface.

Polarimetric imaging is very sensitive to lighting conditions. To simplify the equations involved, it is usually assumed that incoming light radiation is unpolarized (true only on overcast days), to have an estimate of the light subsurface scattering (upwelling) of the sea and to work with cameras geometrically behaving with an orthographic model (reasonable only for very long focal lengths). For all such limitations, this technique is not very popular in acquiring space-time data and is not suitable for a comprehensive freak wave characterization for the limited extent of the captured area.

X-band radars

Marine X-band radars use a lower frequency portion of the electromagnetic spectrum than the optical stereo or polarization-based solutions described above. They work by rotating antenna transmitting pulses of microwaves with a frequency of 8.0–12.0 GHz (wavelength range of 2.5–3.75 cm) at a grazing angle to the sea surface. Assuming that the sea surface exhibits periodic structures with wavelengths in the same range of the emitted electromagnetic radiation (typical for wind-generated capillary waves), the interaction is subject to the so-called Bragg scattering reinforcing the echo signal in the direction of the radar receiver depending on the underlying pattern structure. This causes the captured radar backscatter to be affected by (1) the local slope of the surface, similar to what happens with polarimetric imaging, (2) micro-breakers localized on wave crests, and (3) long-wave orbital motion modulating the capillary surface waves. With Fourier spectral analysis, radar data can be processed to estimate the directional wave spectra of an area of a few squared kilometers. Scaling and Fourier inversion restore the space-time field $\eta(x, y, t)$, in most cases, however, retaining only the linear part of the wave process (Nieto Borge et al., 2004). Despite the large sea area that is recorded with marine radars (order of 1 km $\times$ 1 km), the detection of freak waves is hindered by the temporal (larger than 1 second, a typical antenna rotation period) and spatial (order of meters) resolutions of the system, leading to a systematic underestimation of the highest waves of the field.

Space-time records with freak wave occurrence

Catalogs of measured rogue waves over the global oceans exist in the literature and report the results of multiyear efforts. Most of the data are based on observations from single-point instruments; on the other hand, as for space-time wave data, there is still a scarcity of available records (see, e.g., Guimarães et al., 2020). The complexity of deployment and the computational effort required for processing the data have been limited at the time of writing the use of such a type of observation, hindering long-term measurements for investigating space-time freak waves. Progresses have been recently made, for instance, for the real-time processing of stereo images (Pistellato et al., 2021).

Examples of space-time rogue waves collected in different conditions are given below, produced by using stereo imaging systems. Some details are provided in the figure captions or can be found in the references. All waves are classified as freak meeting the crest criterion ($\eta_{max} \geq 1.25H_s$) for seconds before and after the peak for which the 2-D elevation map is drawn.

Fig. 4.10 shows a freak wave recorded in the Northwestern Pacific during the tropical storm Kong-rey (2018), with winds producing a mixed sea state, a combination of wind sea and swell. The 2-D region R is relatively small compared to the dominant wavelength, with less than a peak wave captured by the cameras. A single individual wave is recorded as having $\eta_{max} = 1.29H_s$, and the temporal profile is symmetric around the maximum elevation. A freak wave peaking at $\eta_{max} = 1.48H_s$ is shown in Fig. 4.11. In this case, the sea state was generated by the passage of an atmospheric front in the Yellow Sea and the

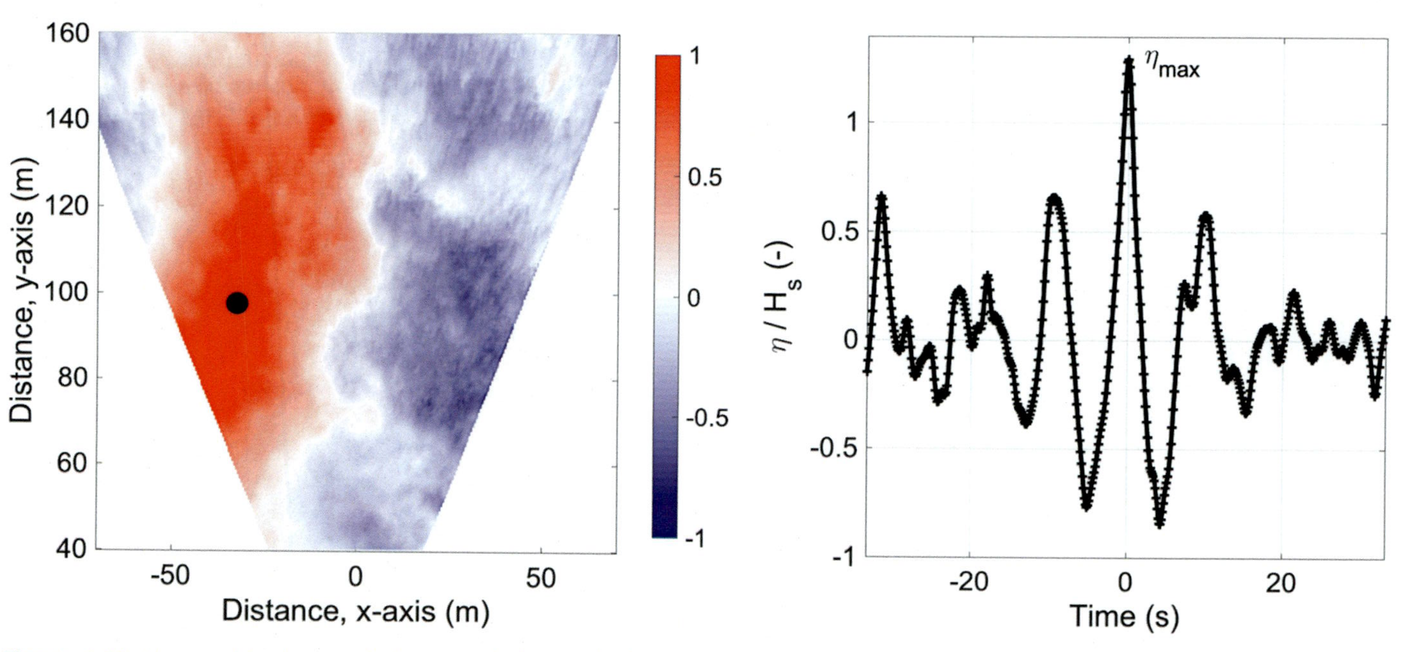

Figure 4.10 Measured freak wave during a tropical storm in the Northwestern Pacific (Benetazzo et al., 2021). Freak wave detection is based on the space-time maximum crest height criterion, with $\eta_{max} = 1.29H_s$. (*left*) Spatial shape $\eta(x, y)$ of the freak wave (P_{freak} is shown with a circle black marker). (*right*) Temporal shape $\eta(t)$ of the freak wave.

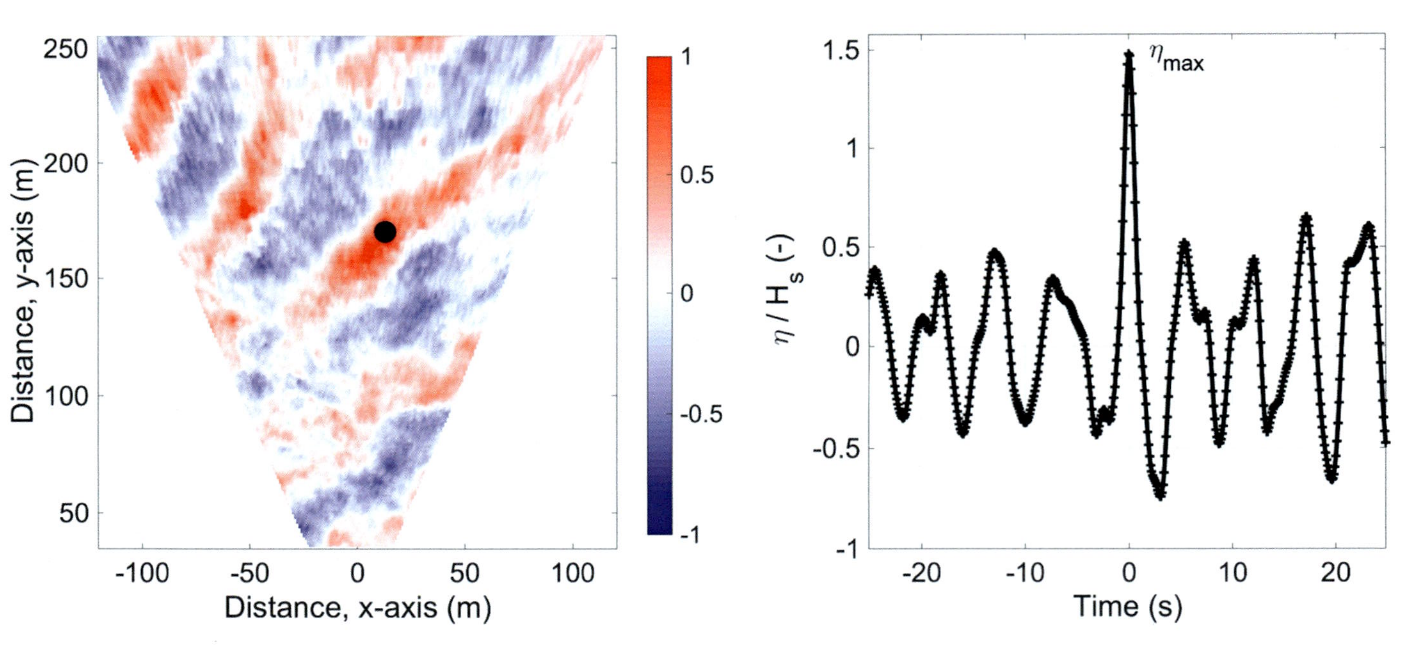

Figure 4.11 Measured freak wave during the passage of an atmospheric front (Benetazzo et al., 2018). Freak wave detection is based on the space-time maximum crest height criterion, with $\eta_{max} = 1.48H_s$. (*left*) Spatial shape $\eta(x, y)$ of the freak wave (P_{freak} is shown with a circle black marker). (*right*) Temporal shape $\eta(t)$ of the freak wave.

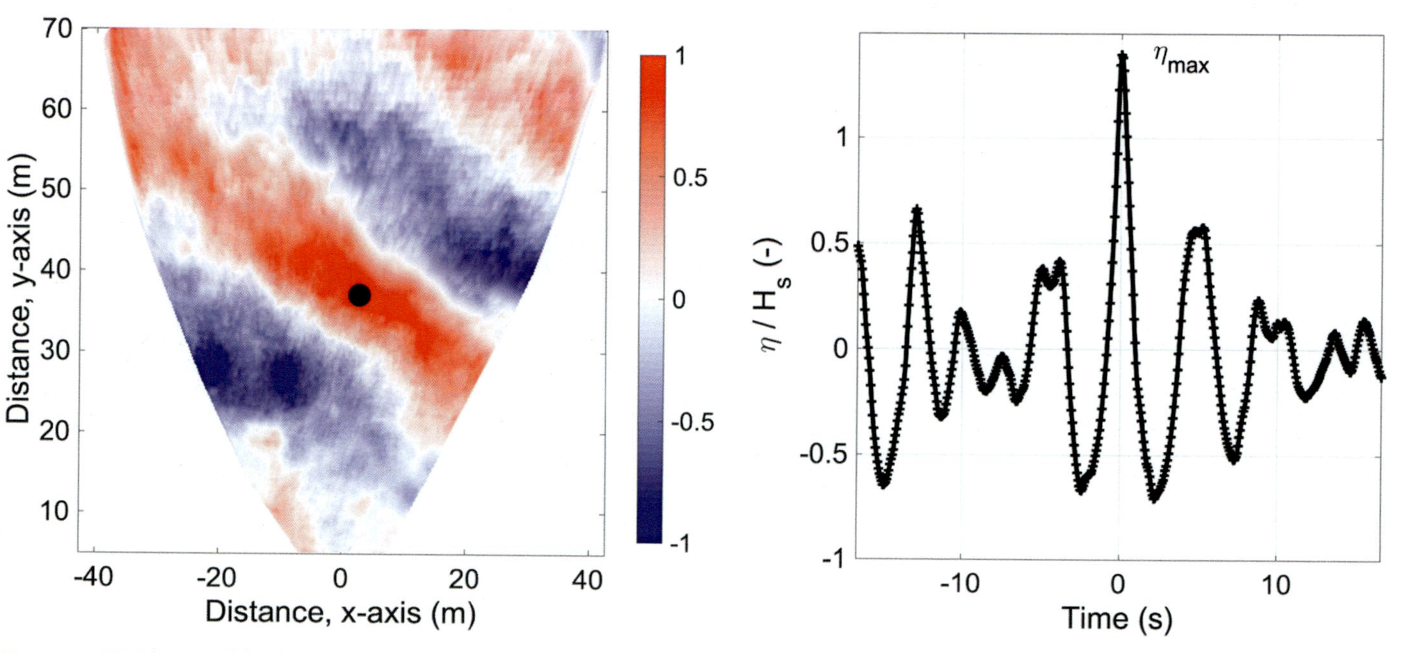

Figure 4.12 Measured freak wave during a unimodal sea state (Benetazzo et al., 2015). Freak wave detection is based on the space-time maximum crest height criterion, with $\eta_{max} = 1.40H_s$. (*left*) Spatial shape $\eta(x, y)$ of the freak wave (P_{freak} is shown with a circle black marker). (*right*) Temporal shape $\eta(t)$ of the freak wave.

2-D surface region wide enough to encompass a few wave and crest lengths. Finally, Fig. 4.12 reports a freak wave event $\eta_{max} = 1.40H_s$ recorded in the North Adriatic Sea (Italy) during a mid-latitude, mild severity storm.

Conclusions

Considerable progress has been made in recent years, both in developing sensors and methods for space-time measurements of oceanic waves and improving knowledge of freak wave amplitude and probability. Specifically, dedicated instruments permit sampling at scales of wave processes leading to the focusing of energy and formation of very high waves compared to those characteristics of the sea state. Stereo imaging is the most comprehensive system used to determine the evolution of wave fields, with some limitations, such as the requirement of natural light conditions and the area of the sea surface recorded that is up to 200×200 m^2.

Improved observation and knowledge of the appearance, dynamics, size, and probability of occurrence of space-time freak waves are crucial to constrain the design of all marine structures (such as ships or offshore platforms) that occupy a surface area $A > 0$. Moreover, toward a more realistic representation of the highest waves that can occur during marine storms, space-time observations of extreme and freak waves provide validation data for numerical models (see Chapter 9).

Looking forward, development efforts directed at improving the use of multi-sensor platforms, fusing imaging systems, radars, and point probes would have an impact on freak wave characterization. In addition, efforts should be invested to broaden access to facilities, technology, and data sets by improving ease of use and making hardware and expertise more readily available.

References

Benetazzo, A., Ardhuin, F., Bergamasco, F., Cavaleri, L., Guimarães, P.V., Schwendeman, M., et al., 2017. On the shape and likelihood of oceanic rogue waves. Scientific Reports 7 (8276), 1–11. Available from: https://doi.org/10.1038/s41598-017-07704-9.

Benetazzo, A., Barbariol, F., Bergamasco, F., Torsello, A., Carniel, S., Sclavo, M., 2015. Observation of extreme sea waves in a space-time ensemble. Journal of Physical Oceanography 45 (9), 2261–2275. Available from: https://doi.org/10.1175/JPO-D-15-0017.1.

Benetazzo, A., Bergamasco, F., Yoo, J., Cavaleri, L., Kim, S.-S., Bertotti, L., et al., 2018. Characterizing the signature of a spatio-temporal wind wave field. Ocean Modelling 129, 104–123. Available from: https://doi.org/10.1016/j.ocemod.2018.06.007.

Benetazzo, A., Barbariol, F., Bergamasco, F., Bertotti, L., Yoo, J., Shim, J.S., et al., 2021. On the extreme value statistics of spatio-temporal maximum sea waves under cyclone winds. Progress in Oceanography 197, 102642. Available from: https://doi.org/10.1016/j.pocean.2021.102642.

Bergamasco, F., Torsello, A., Sclavo, M., Barbariol, F., Benetazzo, A., 2017. WASS: An open-source pipeline for 3D stereo reconstruction of ocean waves. Computers and Geosciences 107, 28–36. Available from: https://doi.org/10.1016/j.cageo.2017.07.001.

Boccotti, P., 2000. Wave Mechanics for Ocean Engineering. Oxford, p. 496.

Fedele, F., 2012. Space–time extremes in short-crested storm seas. Journal of Physical Oceanography 42, 1601–1615. Available from: https://doi.org/10.1175/JPO-D-11-0179.1.

Fedele, F., Benetazzo, A., Gallego, G., Shih, P.-C., Yezzi, A., Barbariol, F., et al., 2013. Space–time measurements of oceanic sea states. Ocean Modelling 70, 103–115. Available from: https://doi.org/10.1016/j.ocemod.2013.01.001.

Fedele, F., Brennan, J., Ponce de Leon, S., Dudley, J., Dias, F., 2016. Real world ocean rogue waves explained without the modulation instability. Scientific Reports. Available from: https://doi.org/10.1038/srep27715.

Fedele, F., Lugni, C., Chawla, A., 2017. The sinking of the El Faro: predicting real world rogue waves during Hurricane Joaquin. Scientific Reports 7, 1118. Available from: https://doi.org/10.1038/s41598-017-11505-5.

Forristall, G.Z., 2006. Maximum wave heights over an area and the air gap problem. In: Proceedings of the ASME 25th International Conference on Offshore Mechanics and Arctic Engineering, Hamburg, OMAE2006–92022, pp. 11–15.

Guimarães, P.V., Ardhuin, F., Bergamasco, F., Leckler, F., Filipot, J.F., Shim, J.S., et al., 2020. A data set of sea surface stereo images to resolve space-time wave fields. Scientific Data 7 (145), 1–12. Available from: https://doi.org/10.1038/s41597-020-0492-9.

Khorram, S., Koch, F.H., van der Wiele, C.F., Nelson, S.A., 2012. Remote Sensing. Springer Science & Business Media.

Malila, M.P., Barbariol, F., Benetazzo, A., Breivik, Ø., Magnusson, A.K., Thomson, J., et al., 2023. Statistical and dynamical characteristics of extreme wave crests assessed with field measurements from the North Sea. Journal of Physical Oceanography 53 (2), 509–531.

Nieto Borge, J.C., Rodriguez, G.R., Hessner, K., Gonzalez, P.I., 2004. Inversion of marine radar images for surface wave analysis. Journal of Atmospheric and Oceanic Technology 21, 1291–1300.

Pezzaniti, J.L., Chenault, D., Roche, M., Reinhardt, J., Schultz, H., 2009. Wave slope measurement using imaging polarimetry. In: Ocean Sensing and Monitoring, 7317. SPIE, pp. 60–72.

Pistellato, M., Bergamasco, F., Torsello, A., Barbariol, F., Yoo, J., Jeong, J.-Y., et al., 2021. A physics-driven CNN model for real-time sea waves 3D reconstruction. Remote Sensing 13, 3780. Available from: https://doi.org/10.3390/rs13183780.

Scherr, T., 2017. Gradient-Based Surface Reconstruction and the Application to Wind Waves (Doctoral dissertation). Available from: https://doi.org/10.11588/heidok.00023653.

Tayfun, M.A., Fedele, F., 2007. Wave-height distributions and non-linear effects. Ocean Engineering, 34, 1631–1649. Available from: https://doi.org/10.1016/j.oceaneng.2006.11.006.

Tayfun, M.A., 1980. Narrow-band nonlinear sea waves. Journal of Geophysical Research 85, 1548–1552. Available from: https://doi.org/10.1029/JC085iC03p01548.

Zappa, C.J., Banner, M.L., Schultz, H., Corrada-Emmanuel, A., Wolff, L.B., Yalcin, J., 2008. Retrieval of short ocean wave slope using polarimetric imaging. Measurement Science and Technology 19 (5), 055503.

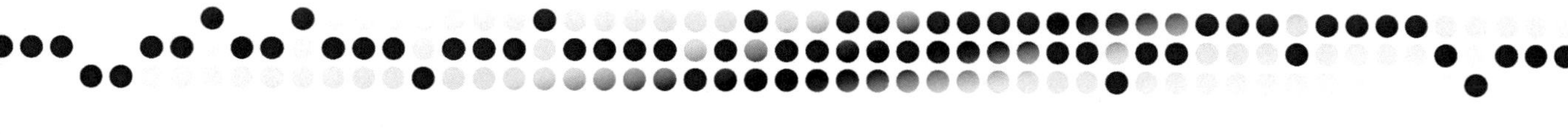

Mechanism 1: nonlinear wave interactions

Nobuhito Mori
Disaster Prevention Research Institute, Kyoto University, Japan

Introduction

Freak/rogue waves (denotes rogue wave hereafter) are high waves that suddenly (beyond expectation) appear in the open ocean. Rogue waves are often defined as the maximum wave that exceeds twice the significant wave height. The study of rogue waves was actively discussed in the engineering community in the first half of the 1990s. It was reported that the cause of the occurrence of rogue waves in the open ocean, in general, was largely due to third- or higher-order nonlinear interactions according to numerical calculations by the high-order spectral (HOS) method (Mori, 2004; Yasuda et al., 1992) and tank experiments (Stansberg, 1992). It began to attract attention from science and engineering fields, and several interdisciplinary international conferences were held around 2000 (Olagnon and Athanassoulis, 2001; Olagnon, 2005).

From the early stage of rogue wave research, it has been suspected that the instability of the wave field caused by third-order nonlinear interference, as typified by the Benjamin–Feir instability, is related to rogue waves due to the similarity with the waveform (Fig. 5.1). The third-order nonlinear interactions increase the generation of rogue waves for a unidirectional wave train when the effects of ocean currents and topography are negligible (Olagnon, 2005). On the other hand, additional studies show that directional dispersion reduces the effects of third-order nonlinear enhancement on rogue waves (Onorato et al., 2009).

This chapter summarizes the rogue research from the three viewpoints of nonlinear wave dynamics, wave statistics, and rogue waves for general single-wave systems in deep water.

Science and Engineering of Freak Waves.
DOI: https://doi.org/10.1016/B978-0-323-91736-0.00002-X

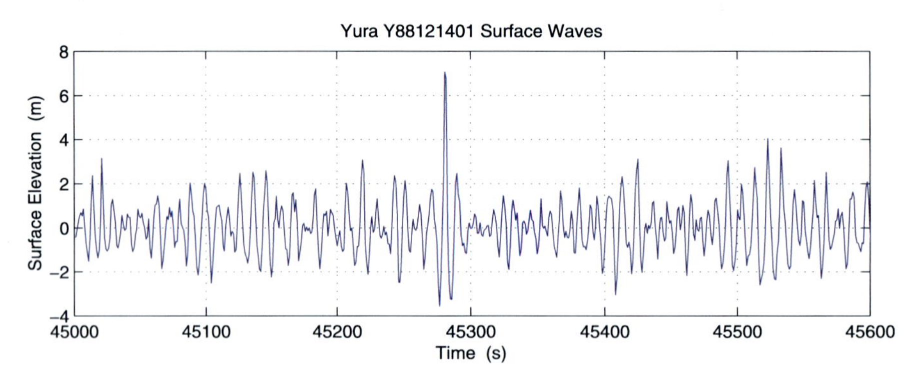

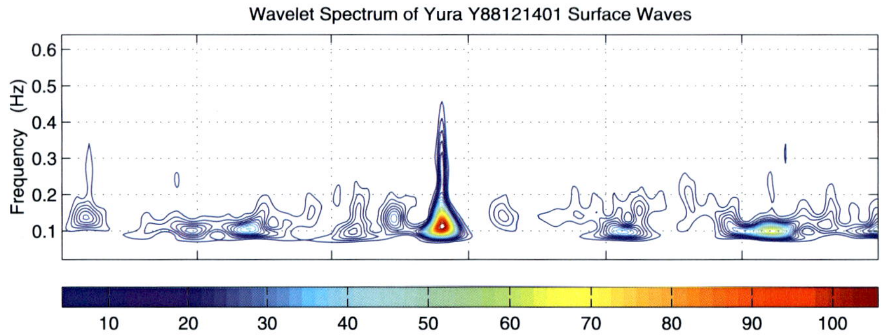

Figure 5.1 Observed rogue wave profile and its wavelet spectra in the Sea of Japan (Mori et al., 2002). *Source: Reprinted with permission from Ocean Engineering, Volume 29 © 2002 by Elsevier.*

Linear random wave theory

Rogue waves are widely defined as waves larger than twice a significant wave height in a wave train. Following this definition, rogue waves can be caused by a random train's linear superposition. As explained in several previous chapters, nonlinear wave interactions can enhance the occurrence of rogue waves. However, it is important to understand the statistical distribution of linear random waves as it relates to nonlinear waves, as explained in Chapter 1.

Assuming a linear, unidirectional wave having a narrowbanded spectrum (hereafter referred to as linear waves), the probability density distribution (PDF) of the water surface displacement η and wave height H is expressed by the normal and Rayleigh distributions, respectively (Rice, 1945).

$$p(H) = \frac{1}{4} H e^{-\frac{1}{8}H^2} \tag{5.1}$$

where H is the wave height normalized by the root mean square (rms) value of the water surface displacement. The phase of each spectral component is assumed to be random. Random-phase approximation is important for both linear and nonlinear wave trains because not only the amplitude but also the phase distribution of the spectral components determines the degree of randomness.

If we regard the rogue wave as the maximum wave in the wave train, the occurrence probability of the rogue wave can be regarded as an estimation of maximum wave height $H_{\max}$ in the wave train. The occurrence probability of $H_{\max}$ depends on the number of waves (or storm duration for a given wave period). For a linear wave case, if the number of waves N in the wave train is constant, the PDF of the maximum wave height $H_{\max}$ can be expressed by

$$p_m(H_{max})dH_{max} = \frac{N}{4} H_{max}\xi \exp(-N\xi)dH_{max} \tag{5.2}$$

$$\xi = e^{-\frac{H_{max}^2}{8}} \tag{5.3}$$

where H_{max} is the wave height normalized by the rms value of the water surface displacement (Goda, 2010). Integrating $H_{max} \geq 8\eta_{rms}(=2H_{1/3})$ in Eq. (5.2), the frequency of rogue wave for a linear narrowband irregular wave P_{freak} is obtained.

$$P_{freak} = 1 - \exp\left(-e^{-8}N\right) \tag{5.4}$$

P_{freak} only depends on the number of waves per observation N.

The observed waves in the ocean slightly differ from Eqs. (5.1) and (5.2). This is because the observed ocean wave height distribution in the sea is less likely to produce high waves than the Rayleigh distribution. This is because waves in the actual ocean have a slightly wider spectral shape than assumed in the Rayleigh distribution.

Nonlinear random wave theory

Unidirectional waves

The nonlinear energy transport by the resonant interactions starts to occur in the third-order nonlinearity. However, Janssen (2003) formulated a theory of homogeneous four-wave interactions, including effects of quasi-resonant transfer, and showed enhancement of rogue wave probability due to the third-order nonlinear interactions (i.e., modulational instability). Modulational instability is a quasi-resonant interaction process. For example, wave numbers and frequencies satisfy the following conditions:

$$\overrightarrow{k_1} + \overrightarrow{k_2} - \overrightarrow{k_3} - \overrightarrow{k_4} = 0 \tag{5.5}$$

$$\omega(\overrightarrow{k_1}) + \omega(\overrightarrow{k_2}) - \omega(\overrightarrow{k_3}) - \omega(\overrightarrow{k_4}) \leq \epsilon^2 \tag{5.6}$$

where k is the wave number, ω is the angular frequency, and $\epsilon = k_0\sqrt{m_0}$ is the representative wave steepness (magnitude of nonlinearity).

Janssen (2003) formulation of changes in kurtosis of the surface elevation μ_4 by the quasi-resonant effects is agreed with the previous research results by numerical and experimental works (Yasuda et al., 1992, Stansberg, 1992). The prediction of rogue waves is closely related to the water surface displacement deviating from the Gaussian process (Mori and Janssen, 2006). The skewness, the third-order moment of the water surface displacement, is proportional to the waveform gradient and is related to the vertical asymmetry of the waveform. Therefore an increase or decrease in skewness significantly affects the amplitude distribution, while it has no effect on wave height because the effect cancels out (Tayfun, 2006). Considering the fourth-order moment of the water surface displacement μ_4, the real term of the nonlinear energy transport function is important. If free waves are considered, the value of μ_4 and the wave action density N have the following relationship (Janssen, 2003).

$$\mu_4 - 3 = \frac{12}{g^2 m_0^2} \int d\vec{k}_{1,2,3,4} T_{1,2,3,4} \sqrt{\omega_1 \omega_2 \omega_3 \omega_4} \delta_{1+2-3-4} R_r(\Delta\omega, t) N_1 N_2 N_3 \tag{5.7}$$

$$R_r = (1 - cos(\Delta\omega t))/\Delta\omega \tag{5.8}$$

where m_0 is the total energy, T is the nonlinear interaction function and R_r is the nonlinear energy transport function. In this case, both resonance and non-resonance contribute to the time evolution of kurtosis. Thus, given the action density $N(\vec{k})$, the nonlinear nuclear function $T_{1,2,3,4}$ determines the change in kurtosis due to the four-wave interaction. Of course, kurtosis also varies with the bound wave component, but this is proportional to the square of wave steepness and independent of spectral shape.

Replacing the action density N to the wavenumber spectrum $F(\vec{k})$ and frequency spectrum $E(\omega, \theta)$, neglecting the effect of bound waves, assuming unidirectional waves and approximate wave spectra,

$$E(\omega) = \frac{m_0}{\sigma_\omega \sqrt{2\pi}} e^{-\frac{1}{2}\nu^2} \tag{5.9}$$

where $\nu = (\omega - \omega_0)/\sigma_\omega$ is the frequency normalized by the peak frequency ω_0 and the spectral width with a small quantity $\Delta = \sigma_\omega/\omega_0$, and Eq. (5.7) can be rewritten by Eq. (5.9) as follows using

$$\kappa_{40} = \frac{24\epsilon^2}{\Delta^2} P \int \frac{d\nu_1 d\nu_2 d\nu_3}{(2\pi)^{3/2}} \frac{e^{-\frac{1}{2}[\nu_1^2 + \nu_2^2 + \nu_3^2]}}{(\nu_1 + \nu_2 - \nu_3)^2 - \nu_1^2 - \nu_2^2 + \nu_3^2} \tag{5.10}$$

where P is the principle value to avoid the singularity of the integral.

Janssen (2003) introduced Benjamin–Feir index (BFI), and Eq. (5.10) is rewritten as a function of excess kurtosis $\mu_4 - 3$

$$\mu_4 - 3 = \frac{\pi}{\sqrt{3}} BFI^2 \tag{5.11}$$

$$BFI = \frac{\epsilon}{\Delta} \sqrt{2} \tag{5.12}$$

As shown by Alber (1978) and Alber and Saffman (1978), the propagation of gravity waves involves both energy concentration due to wave steepness ϵ, and linear dispersion Δ plays an important role in the propagation of waves. The nonlinear enhancement of rogue wave in deep water occurs $BFI > 1$. From Eqs. (5.11) and (5.12), μ_4 is proportional to the wave steepness and the inverse of spectral width. Therefore the nonlinear effect becomes stronger for waves with a large wave steepness and narrowband spectral width.

Eq. (5.11) can be rewritten using Goda's spectral width parameter Q_p.

$$\mu_4 - 3 = \frac{2\pi^2}{\sqrt{3}} k_p^2 m_0 Q_p^2 \propto \frac{H_{1/3}^2 Q_p^2}{T_{1/3}^4} \tag{5.13}$$

Furthermore, assuming a developing wind wave and applying Toba's 3/2 power law, Eq. (5.13) gives

$$\mu_4 \propto Q_p^2 / H_*^{\frac{2}{3}} \tag{5.14}$$

Eq. (5.14) implies that μ_4 decreases the wave height normalized wave height H_* by the friction velocity u_* increases and it conversely increases with the sharpening of the spectral width. Note that there is no tuning parameter in Eq. (5.11), which can easily be calculated from wave spectra in Eq. (5.13).

Multidirectional waves

In the above treatment, the narrowband spectra assumption and directional dispersion of the frequency spectrum were assumed to be negligible. However, the experimental data showed a reduction of nonlinear enhancement to kurtosis and rogue wave by wave directionality (Waseda et al., 2009; Onorato et al., 2009).

Janssen and Bidlot (2009) included directional effects and studied the kurtosis behavior of short-term solutions. For two-dimensional propagation, ω_4 in Eq. (5.6) becomes

$$\begin{aligned} \omega_4 = \Big\{ \left(\omega_1^2 + \omega_2^2 - \omega_3^2\right)^2 &+ 2\omega_1^2\omega_2^2[\cos(\theta_1 - \theta_2) - 1] \\ &- 2\omega_1^2\omega_3^2[\cos(\theta_1 - \theta_3) - 1] \\ &- 2\omega_2^2\omega_3^2[\cos(\theta_2 - \theta_3) - 1] \Big\}^{1/4} \end{aligned} \tag{5.15}$$

Denoting the width of the directional spectrum by σ_θ, we can normalize the angular frequency and directional widths as

$$\delta_\omega = \frac{\sigma_\omega}{\omega_0} \tag{5.16}$$

$$\delta_\theta = \sigma_\theta \tag{5.17}$$

and in the narrowband spectra approximation, they are assumed to be small. Expanding ω_4 in the small parameters δ_ω and δ_θ, the frequency mismatch $\Delta\omega$ in Eq. (5.7) becomes

$$\Delta\omega = \delta_\omega^2 \omega_0 \{(\nu_3 - \nu_1)(\nu_3 - \nu_2) - R(\phi_3 - \phi_1)(\phi_3 - \phi_2)\} + O(\delta^3) \tag{5.18}$$

where ϕ is the normalized direction parameter as $\phi = (\theta - \theta_0)/\sigma_\theta$ by the directional width σ_θ, and R is a parameter that is a measure of the angular width concerning the frequency width

$$R = \frac{1}{2}\frac{\delta_\theta^2}{\delta_\omega^2}\left(= \frac{1}{2}\omega_0^2 \frac{\sigma_\theta^2}{\sigma_\omega^2} \right) \tag{5.19}$$

In the narrowband spectra approximation, the transfer coefficient $T_{1,2,3,4}$ in Eq. (5.7) can be approximated by k_0^3. Janssen and Bidlot (2003) studied both the short-time and long-time behavior of kurtosis by substituting Eq. (5.18) to Eq. (5.7). For the short timescale, a result is obtained, which holds for general spectra. The kurtosis becomes

$$\mu_4 - 3 = 3\tau^2 BFI^2 (1 - R) \tag{5.20}$$

where $\tau = \omega_0 \delta_\omega^2 t$ is a dimensionless time from the initial condition. This result shows that directional effects play an important role. Both positive and negative kurtosis can occur. The positive case corresponds to nonlinear focusing when $\delta_\theta < \sqrt{2}\delta_\omega$ while for narrower directional spread ($\delta_\theta < \sqrt{2}\delta_\omega$). The condition $R = 1 \rightarrow \sigma_\theta = \sqrt{2}\delta_\omega$ corresponds with one of the boundaries of the instability diagram of the two-dimensional nonlinear Schrödinger equation (NLS) (Alber and Saffman, 1978).

Based on the Monte Carlo simulation using the two-dimensional nonlinear Schrödinger equation (MC-CNLS), Mori et al. (2011) parameterized kurtosis as a function of *BFI* and frequency and directional ratio *R*.

$$\mu_4 - 3 = \frac{\pi}{\sqrt{3}} BFI_{2D}^2 \tag{5.21}$$

$$BFI_{2D}^2 = \frac{BFI^2}{1 + \alpha_2 R} \tag{5.22}$$

where α_2 is the empirical tuning coefficient by MC-CNLS and becomes 7.10.

Fig. 5.2 shows the relationship between μ_4, BFI, and σ_θ (Mori et al., 2011). The magnitude of kurtosis depends on the ratio of directional dispersion and frequency dispersion. The narrower the directional spectra, the larger the kurtosis (i.e., the frequency of rogue waves is enhanced).

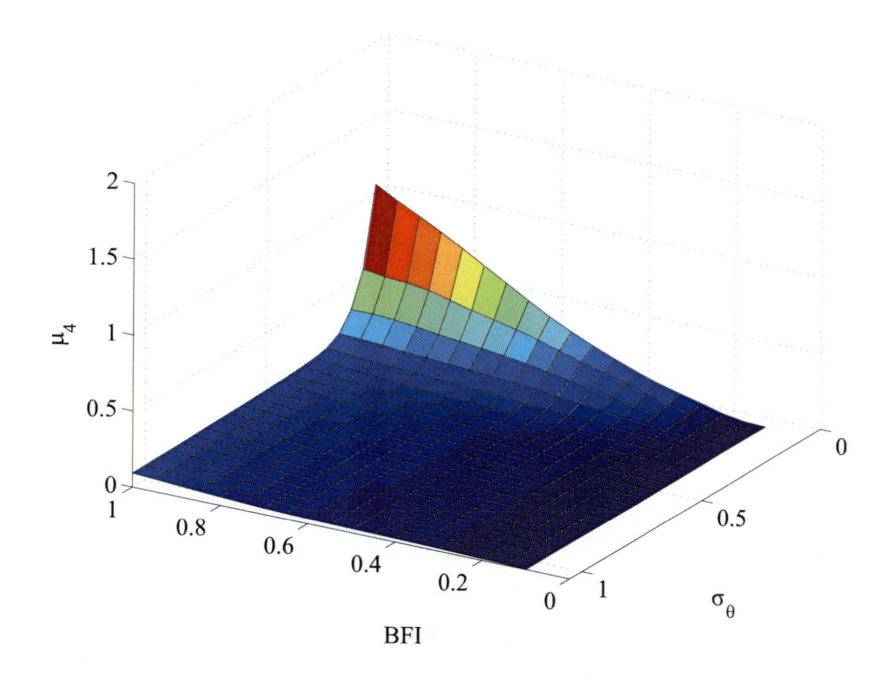

Figure 5.2 Relation between μ_4, BFI, and σ_θ. *Source: Revised from Mori, N., Onorato, M., Janssen, P.A. 2011. On the estimation of the kurtosis in directional sea states for freak wave forecasting. Journal of Physical Oceanography 41(8), 1484–1497. Reprinted with permission from* Journal of Physical Oceanography *© 2012 by American Meteorological Society.*

Nonlinear wave height statistics

The third-order nonlinear interactions change kurtosis and related rogue waves. We need to predict for the occurrence probability of rogue waves as the extension of linear narrowbanded wave theory, as shown in Eqs. (5.1) to (5.4). How does the frequency of the occurrence of rogue waves change when kurtosis is changed from the Gaussian process. First, let us calculate the water surface displacement $\eta(t)$ and its auxiliary variables $\zeta(t)$ normalized by variance σ

$$Z(t) = \eta(t) + i\zeta(t) = A(t)e^{i\phi(t)} \tag{5.23}$$

where A and ϕ denote the amplitude and phase of the wave envelope. Assuming the PDF of the water surface displacement deviates slightly from the central limit theorem and has an Edgeworth distribution. If the η and ζ are uncorrelated, the following joint PDF is derived.

$$p(\eta, \zeta) = \frac{1}{2\pi}\exp\left[-\frac{1}{2}(\eta^2 + \zeta^2)\right]\left[1 + \frac{1}{3!}\sum_{n=0}^{3}\frac{3!}{(3-n)!n!}\kappa_{(3-n)n}H_{3-n}(\eta)H_n(\zeta) + \frac{1}{4!}\sum_{n=0}^{4}\frac{4!}{(4-n)!n!}\kappa_{(4-n)n}H_{4-n}(\eta)H_n(\zeta)\right] \tag{5.24}$$

where H_n is the nth-order Hermite polynomial, κ is the cumulant, and all variables are nondimensionalized by the rms value of η. Converting η and ζ into A and ϕ and integrating the phase ϕ with $[0, 2\pi]$, the PDF of envelope amplitude A is obtained as similar to Rayleigh distribution of Eq. (5.1). Expanding Eq. (5.24) to the third-order terms and assuming $\kappa_{30} \simeq 0$, we obtain a PDF for the wave height H as twice the amplitude A

$$p(H) = \frac{1}{4} He^{-\frac{1}{8}H^2}[1 + \kappa_{40} A_H(H)] \tag{5.25}$$

$$A_H(H) = \frac{1}{384}\left(H^4 - 32H^2 + 128\right) \tag{5.26}$$

where κ_{40} is the fourth-order cumulant and is equal to $\mu_4 - 3$.

Similar to Eq. (5.2), the distribution of maximum wave height for the weakly nonlinear field can be obtained.

$$p_m(H_{max})dH_{max} = \frac{N}{4} H_{max}\xi \exp(-N\xi)dH_{max} \tag{5.27}$$

$$\xi = e^{-\frac{H_{max}^2}{8}}\left[1 + (\mu_4 - 3)B_H(H_{max})\right] \tag{5.28}$$

From Eq. (5.27), the distribution of maximum wave heights is given by the number of waves N and the kurtosis. Furthermore, the occurrence probability of the rogue wave for the weakly nonlinear field P_{freak} can be obtained.

$$P_{freak} = 1 - \exp\left[-e^{-8}N(1 + 8(\mu_4 - 3))\right] \tag{5.29}$$

Note that the first order of nonlinear contribution to the wave height is the kurtosis because the skewness contribution is canceled out. On the other hand, skewness contribution is the first-order nonlinear contribution to wave crest distribution (Tayfun, 2006).

Eqs. (5.25), (5.27) and (5.29) agree with the linear wave theory based on the Rayleigh distribution with $\mu_4 = 3$. From Eq. (5.29), the effect of nonlinearity on the frequency of occurrence of rogue waves is sufficiently large compared to the linear theory (e.g., $\mu_4 = 3.125$). As long as we consider the maximum wave height in a wave train, kurtosis and the number of waves are important parameters. To verify this result, comparing the results with conditionally sampled data for the number of waves and kurtosis is necessary. It will be discussed in the next section.

Discussion

Estimating the frequency of rogue waves is equivalent to theoretically estimating the maximum wave height. In other words, the essence of predicting rogue waves is to estimate the shape of the maximum wave height distribution. In order to verify the accuracy of the framework for the estimation of the maximum wave height under consideration of the weak wave nonlinearity, it is necessary to explore a large number of waves, different steepness, and frequency bandwidth in the parameter space of the number of waves N and the relative nonlinearity *BFI* (and kurtosis). The above framework is a prediction of rogue waves for given initial conditions. Therefore validation can be possible through experiments.

Many experimental studies on the evolution of random waves discuss the nonlinear enhancement of rogue wave probability. One example is large ensemble experiments for unidirectional wave trains using a large channel (length 270 m $\times$ width 10.6 m $\times$ maximum water depth 10 m) by Mori et al. (2007). In order to compare with theory, more than 10,000 waves were generated, changing *BFI*. In addition to the experiments, the buoy observation data at 30 m depth on the Pacific Ocean side were also used for comparison. The experimental and observational data were conditionally sampled to produce a sample data set of the maximum wave heights, in which both wave number and nonlinearity were classified. Fig. 5.3 compares the PDF of maximum wave heights between the theory (Eq. 5.27), experimental data, and observed data for given N and kurtosis. The number of waves N is fixed at 150 here. The linear wave theory and Rayleigh theory Eq. (5.2) are plotted together. The peak of PDF by the experimental and field data distribution shifts to a large value compared to the linear theory. The nonlinear wave theory Eq. (5.27) shows good agreement with the data for the large kurtosis case, as shown in Fig. 5.3B, although the nearly Gaussian case ($\mu_4 = 3.1$) shows slight overestimation due to spectral bandwidth.

Regarding validation of the occurrence probability of rogue waves, Fig. 5.4 shows the relation between the occurrence probability and μ_4 in different experimental conditions. The solid and dashed lines in the figure show the frequency of the nonlinear and linear theories (Eqs. 5.4 and 5.29). The frequency of occurrence of the rogue waves increases linearly as the value of μ_4 and the experimental results can be explained by the nonlinear theory in Eq. (5.29), although the nonlinear theory slightly exceeds in the region of $\mu_4 < 3.75$. There are several reasons why the nonlinear theory is slightly overestimated compared to the experimental results. The most significant source is the assumption of narrowbanded spectra based on the envelope theory. Although no turning is applied to Eq. (5.29), this kind of error can be optimized based on the data.

Estimating the occurrence probability of rogue waves by BFI and nonlinear wave statistical theory is good for unidirectional waves. However, Waseda et al. (2009) discussed the reduction of nonlinear enhancement by directional dispersion. They found that larger directionality suppresses increased kurtosis. Onorato et al. (2009) summarized two different series of experimental data for

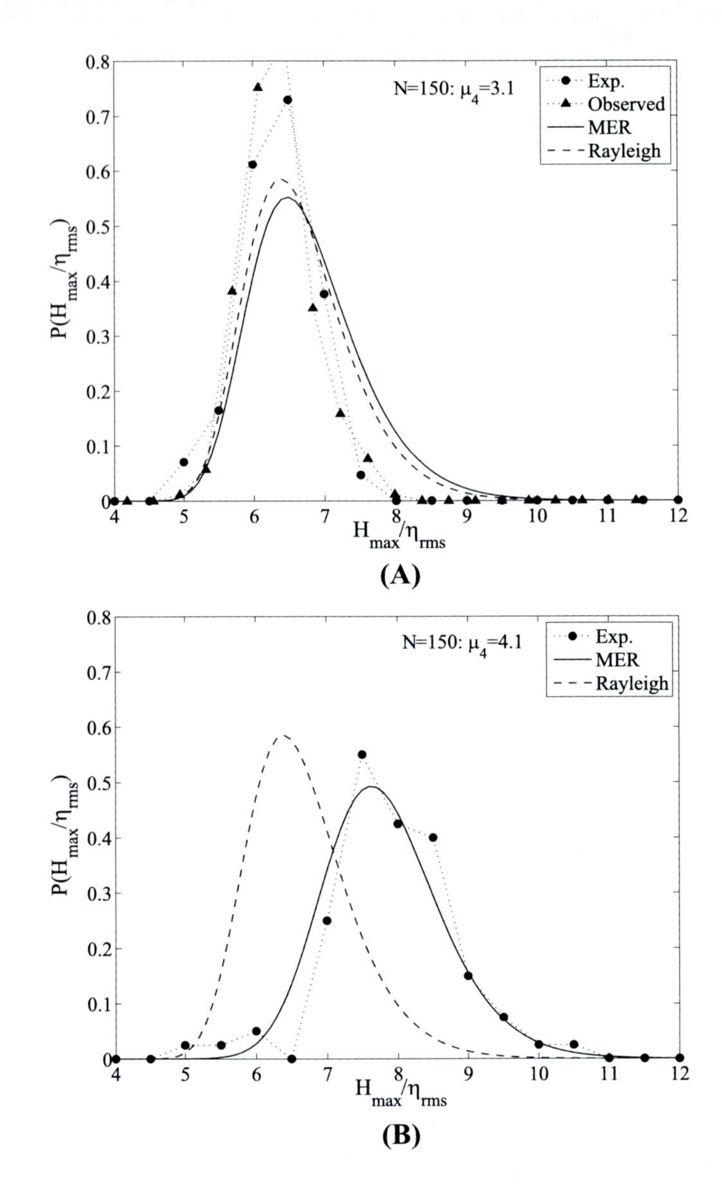

Figure 5.3 Effect of μ_4 in the maximum wave height distribution: $N = 150$ [●: experimental results, ▲: observed data, solid line: Eq. (5.27), dotted line: Eq. (5.2)] (Mori et al., 2007). *Source: Reprinted with permission from* Journal of Geophysical Research *© 2012 by American Geophysical Union.*

validation of maximum kurtosis estimation as a function of *BFI* and directional spreading. Mori et al. (2011) compared the mean value of the kurtosis along the tank for different directional spreading from the two different wave experiments, MCNLS and Eq. (5.29), as shown in Fig. 5.5. As shown in Fig. 5.5 for

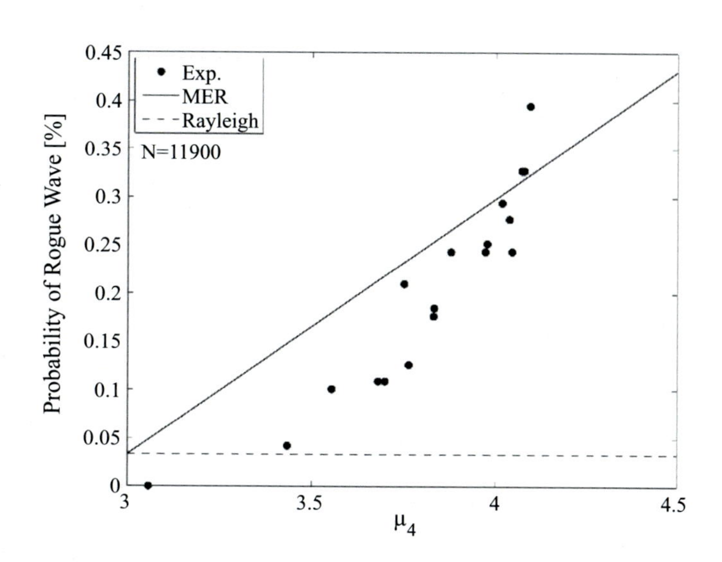

Figure 5.4 Rogue wave occurrence probability and μ_4 in the wave flume data: $N = 11,900$ [●: experimental results, solid line: Eq. (5.29), dotted line: Eq. (5.5)] (Mori et al., 2007). *Source: Reprinted with permission from* Journal of Geophysical Research *© 2012 by American Geophysical Union.*

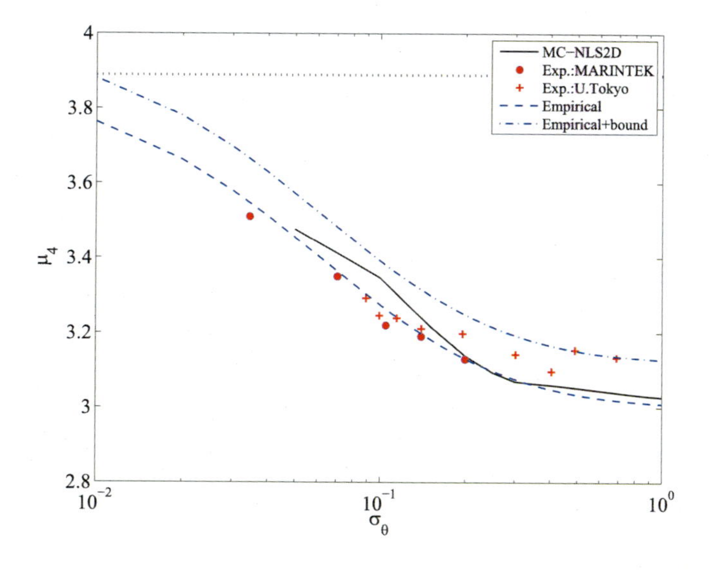

Figure 5.5 Comparison with experimental data: $BFI = 0.7$ (solid line: MCNLS2D, broken line: multidirectional *BFI2D* plus bound mode effects, dotted line: unidirectional *BFI*, ● and +: experimental data) (Mori et al., 2011). *Source: Reprinted with permission from* Journal of Physical Oceanography © *2012 by American Meteorological Society.*

$BFI = 0.7$, the kurtosis remains more or less constant for $\sigma_\theta > 0.2$ and then kurtosis increases as σ_θ decreases. Extreme waves are more probable in the case of narrowbanded directional waves. Overall, the MCNLS and Eq. (5.29) agree with the experimental data. Therefore the empirical formula of maximum kurtosis estimation, including directional dispersion Eq. (5.29), works well for directional waves in a stationary condition such as a wave tank experiment.

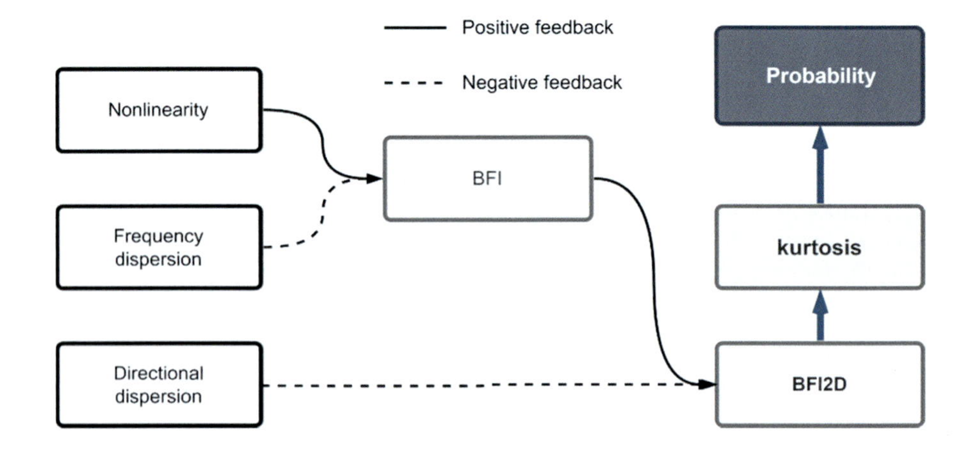

Figure 5.6 Summary of estimating rogue wave probability from given initial conditions.

Summary

Research on rogue waves has been conducted in various fields, from nonlinear wave dynamics to nonlinear statistics and numerical wave forecasting systems (Chapters 8, 9, and the others). In this chapter, we explained the fundamental research flow of rogue wave prediction from the viewpoints of both nonlinear wave dynamics and wave statistics targeting to estimate a maximum wave height in a wave train. The semiempirical equations for predicting kurtosis, the fourth-order moment of water surface displacement, and the effect of directional dispersion are summarized. The results show that, at least for the 2D cross-sectional tank experiments results, it is possible to estimate the distribution of maximum wave height from the directional spectra. Fig. 5.6 summarizes the flowchart to estimate the probability of rogue waves based on the nonlinear theory given initial conditions.

Future research should include verification of kurtosis estimation in multidirectional, multimodal fields and bathymetric changes (Chapters 6 and 7). Recently, Liu et al. (2022) summarized a comprehensive study on crossing sea states using a HOS method to investigate the nonlinear statistics and rogue wave occurrences. Additionally, an experiment is an initial value problem, but there are no initial values in the real ocean. A new approach that links these theories is needed for forecasting in the real sea.

References

Alber, I., 1978. The effects of randomness of the stability of two-dimensional surface wave-trains. Proceedings of the Royal Society London 363A, 525–546.

Alber, I., Saffman, P., 1978. Stability of random nonlinear deep-water waves with finite band-width spectra. Technical Report 31326-6035-RU-00, TRW Defense and Space System Group.

Goda, Y., 2010. Random Seas and Design of Maritime Structures, vol. 33. World Scientific Publishing Company.

Janssen, P.A.E.M., Bidlot, J.R., 2009. On the extension of the freak wave warning system and its verification. European Centre for Medium-Range Weather Forecasts, Reading, UK, p. 42.

Janssen, P.A., 2003. Nonlinear four-wave interactions and freak waves. Journal of Physical Oceanography 33 (4), 863–884.

Janssen, P.A.E.M., Bidlot, J., 2003. New wave parameters to characterize freak wave conditions. Research Department Memo R60. 9. PJ/0387, ECMWF. Reading, UK.

Liu, S., Waseda, T., Yao, J., Zhang, X., 2022. Statistical properties of surface gravity waves and freak wave occurrence in crossing sea states. Physical Review Fluids 7 (7), 074805.

Mori, N., Liu, P.C., Yasuda, T., 2002. Analysis of freak wave measurements in the Sea of Japan. Ocean Engineering 29 (11), 1399–1414.

Mori, N., 2004. Occurrence probability of freak wave in nonlinear wave field. Ocean Engineering 31 (2), 165–175.

Mori, N., Janssen, P.A., 2006. On kurtosis and occurrence probability of freak waves. Journal of Physical Oceanography 36 (7), 1471–1483.

Mori, N., Onorato, M., Janssen, P.A., Osborne, A.R., Serio, M., 2007. On the extreme statistics of long-crested deep water waves: theory and experiments. Journal of Geophysical Research: Oceans 112 (C9).

Mori, N., Onorato, M., Janssen, P.A., 2011. On the estimation of the kurtosis in directional sea states for freak wave forecasting. Journal of Physical Oceanography 41 (8), 1484–1497.

Olagnon, M., Athanassoulis, G.A., 2001. Rogue Waves 2000: Proceedings of a Workshop, Organized by IFREMER and Held in Brest, France, 29–30 November 2000, Within the Brest SeaTechWeek 2000, vol. 32. Editions Quae.

Olagnon, M., 2005. Rogue Waves 2004. Proceedings of a Workshop Organized by IFREMER and held in Brest, France 20–21–22 October 2004 within the Brest Sea Tech Week 2004.

Onorato, M., Waseda, T., Toffoli, A., Cavaleri, L., Gramstad, O., Janssen, P.A.E.M., et al., 2009. Statistical properties of directional ocean waves: the role of the modulational instability in the formation of extreme events. Physical Review Letters 102 (11), 114502.

Rice, S.O., 1945. Mathematical analysis of random noise. The Bell System Technical Journal 24 (1), 46–156.

Stansberg, C.T., 1992. On spectral instabilities and development of nonlinearities in propagating deep-water wave trains. Coastal Engineering Proceedings (23).

Tayfun, M.A., 2006. Statistics of nonlinear wave crests and groups. Ocean Engineering 33 (11–12), 1589–1622.

Waseda, T., Kinoshita, T., Tamura, H., 2009. Evolution of a random directional wave and freak wave occurrence. Journal of Physical Oceanography 39 (3), 621–639.

Yasuda, T., Mori, N., Ito, K., 1992. Freak waves in unidirectional wave trains and their properties. Coastal Engineering Proceedings (23).

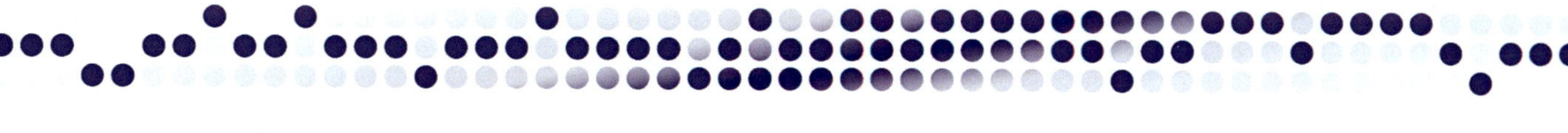

Mechanism 2: crossing waves

Suzana Ilic[1], *Jamie Luxmoore*[1,2] *and Nobuhito Mori*[3]

[1]Lancaster Environment Centre, Lancaster University, Lancaster, United Kingdom [2]Orcina Ltd., Ulverston, United Kingdom [3]Disaster Prevention Research Institute, Kyoto University, Japan

Introduction

One of the phenomena attracting substantial research attention recently is rogue waves in crossing seas, when two or more distinct spectral components are crossing at an angle to each other. Bimodal or crossing seas are a common feature in oceans. They can be observed in general sea states (Semedo et al., 2011) and in extreme seas such as tropical cyclones (Mori, 2012). This bimodal state can also be caused by rapidly turning wind, which will result in two sea states with similar frequency distributions and different directions.

Several marine accidents involving large or rogue waves are known to have occurred in crossing sea conditions (Adcock and Taylor, 2014; Toffoli et al., 2005), including the well-known Draupner or New Year wave (Onorato et al., 2006). McAllister et al. (2019), following earlier work by Adcock and Taylor (2009) and Adcock et al. (2011), showed through a laboratory experiment that the Draupner wave could only have occurred in a crossing sea. Hindcasts consistently provide evidence suggesting that separate wind and swell systems were present at different angles leading to the formation of the Draupner wave (Adcock et al., 2011). Cavaleri et al. (2012) in their study of the Louis Majesty accident demonstrated that rogue waves were formed in crossing sea states with wave systems of similar peak frequency and crossing angles between 40 and 60 degrees. Similarly, Trulsen et al. (2015) showed that an accident to the Prestige tanker took place in a crossing sea state with crossing angles of 90 degrees. Other studies also detected rogue waves in crossing seas (De Pinho et al., 2004; Rosenthal and Lehner, 2008). Recently, Davison et al. (2022) showed that extreme waves (possibly rogue waves) were formed in crossing sea states during a tropical cyclone. Tamura et al. (2009) investigated an incident that occurred in unimodal wave conditions but found that conditions

Science and Engineering of Freak Waves.
DOI: https://doi.org/10.1016/B978-0-323-91736-0.00004-3

suitable for rogue wave formation were formed from a prior crossing sea state. They concluded that a crossing swell and wind sea transformed into a freakish unimodal sea by transfer of energy from the wind sea into the swell mode.

Crossing seas can generate extreme and freak wave events over different timescales (Tamura et al., 2009). The first mechanism is a linear superposition of two systems, which can generate extreme crossing waves continuously over short timescales. The second mechanism is nonlinear interactions between two systems, which can be both the second-order nonresonant (or bound harmonic waves) and the third-order quasi-resonant-type interactions that can generate extreme waves over a long timescale.

This chapter summarizes the mechanisms responsible for the formation of rogue waves in crossing seas, laboratory measurements, numerical simulations of crossing wave fields, and nonlinear wave statistics.

Phenomena and governing equations for describing crossing seas

The physical mechanisms responsible for the formation of rogue waves in crossing seas are still not well understood. Linear and nonlinear processes have been found responsible for generating rogue waves in crossing seas. In the former case, the directional and dispersive interference of two wave systems with differing peak wave directions can lead to much higher wave amplitudes. In the latter case, both second-order and third-order wave nonlinearities are considered to be responsible for the formation of the highest waves.

Second-order nonlinear interaction

A recent analysis of the three simulated sea states based on measurements of the Draupner, Andrea and Killard rogue waves (Fedele et al., 2016) suggests that second-order bound mode effects enhance the directional and dispersive interference that is the main driver for observed rogue wave activity in the ocean. The low-frequency bound waves formed in crossing seas lead to set-up underneath the highest waves, which is opposite to the case in following seas where set-down forms underneath the highest waves. Hence in crossing seas, set-up can increase the crest amplitude of the highest waves. This can be theoretically predicted (Christou et al., 2009) based on second-order interaction kernels. A set-up of the wave-averaged free surface was detected with a wave group in the Draupner wave (Walker et al., 2004) and experimentally shown by McAllister et al. (2019). McAllister et al. (2018) used the multicomponent second-order wave theory for describing the second-order nonlinear interactions that can occur in crossing seas to complement laboratory measurements of set-up in directionally spread and crossing wave groups. They measured and predicted a set-up for wave groups with a Gaussian angular amplitude distribution with standard deviations of above

30–40 degrees (21–28 degrees for energy spectra), and for crossing wave groups with angle of separation of 50–70 degrees and above. Whether the set-up takes place at the point of focus of wave groups depends on the phases of the linear wave groups. They confirmed these results through laboratory experiments.

Third-order interactions

Thus far, much of the data from crossing seas involving rogue wave formation can be adequately explained by second-order theory; however, in some cases, the formation of rogue waves involves rapid changes of the spectrum, primarily as a result of third-order resonant interactions in deep water. In this case, the interaction between wave components satisfied the dispersion relationship. It forces a wave mode that can propagate freely, and energy is exchanged between the different components, which leads to a change in the spectrum (Gibson and Swan, 2007). Waseda et al. (2015) demonstrated this in a laboratory experiment. Two waves at a crossing angle of 9.4 degrees satisfying the dispersion relationship were simulated, resulting in a third wave that formed at 90 degrees and grew as the waves propagated down the tank. The evolution of this wave was explained well with the third-order resonance theory based on Zakharov's reduced gravity equation (Zakharov, 1968).

Furthermore, non-linear interactions between two crossing wave systems can lead to the formation of extreme waves (Davison et al., 2022). However, Trulsen et al. (2015), found no evidence that non-linear interactions between two crossing wave systems influenced the maximum wave height or kurtosis in the Prestige accident. Additional factors such as the energy ratio between the constituent systems, as well as their frequency and angular separation, may also influence the development of extreme and rogue waves (Davison et al., 2022).

Several studies have explored nonlinear instabilities in crossing seas (e.g. Onorato et al., 2006; Ruban, 2010). When unidirectional wave groups or narrow-banded directional wave groups cross, third-order nonlinear quasi-resonant free-wave interactions, and associated modulational instabilities, can occur. Crossing seas may also enhance the probability of the occurrence of rogue waves through modulational instability (Toffoli et al., 2011; Cavaleri et al., 2012). These studies did not consider directional spreading.

Phase-resolving equations for describing crossing seas

Coupled nonlinear Schrödinger equations

The evolution of wave envelopes and, in particular, modulational instability have been widely studied using the nonlinear Schrödinger equation (NLS). However, in order to simulate crossing seas, this equation has to be modified; coupled nonlinear Schrödinger (CNLS) equations have been developed (Onorato et al., 2006, 2010). These were, for example, used to examine the modulational instability of two wave systems with different directions of propagation (Onorato et al., 2010; Shukla et al., 2006).

Onorato et al. (2006) found that introducing a second wave system can increase the instability growth rate and increase the size of the unstable region. Shukla et al. (2006) found that two crossing long-crested wave trains can form large-amplitude wave groups even when the individual wave trains are modulationally stable. Two-wave coupled systems show increased nonlinear focusing and decreased time to develop large waves compared to a non-coupled system (Grönlund et al., 2009). Gramstad and Trulsen (2010) derived the fourth-order CNLS equation and applied it to the more general case of two Stokes waves with different directions of propagation and frequencies. They found that the addition of a swell wave system slightly increases the number of rogue waves in a short-crested wind sea.

There are limitations in using NLS type equations as the instability results can extend beyond the bandwidth constraints of the equations (Gramstad and Trulsen 2011).

Coupled Zakharov equations

The more general equation is the Zakharov equation (Zakharov, 1968), which includes energy exchange among the wave components due to resonant and quasi-resonant four-wave interactions. Unlike the NLS, the Zakharov equation does not have bandwidth constraints, and hence it can be applied to a wider range of crossing sea systems. Gramstad et al. (2018) derived the modulational instability of a system of two Stokes waves using the Zakharov equation (Zakharov, 1968) and the work of Okamura (1984) on standing waves.

Gramstad et al. (2018) included a wave system consisting of two Stokes waves with wave numbers *ka* and *kb* with corresponding complex amplitude functions *A(t)* and *B(t)*, and their associated set of perturbations $ka \pm K$ and $kb \pm K$ into the Zakharov equation. Keeping only linear terms in the perturbation amplitudes and defining amplitudes as a function of the Zakharov kernel functions and an angular frequency associated with perturbations Ω, they obtained a system of equations, which can be written as $Mx = 0$; where $x = [a_+, a_-{}^*, b_+, b_-{}^*]^T$ and

$$M = \begin{bmatrix} M_a^+ + \Omega & -aaT_{a+,a-,a,a} & -2ab^*T_{a+,b,a,b+} & -2abT_{a+,b-,a,b} \\ -a^*a^*T_{a+,a-,a,a} & M_a^- - \Omega & -2a^*b^*T_{a-,b+,a,b} & -2a^*bT_{a-,b,a,b-} \\ -2a^*bT_{a+,b,a,b+} & -2abT_{a-,b+,a,b} & M_b^+ + \Omega & -bbT_{b+,b-,b,b} \\ -2a^*b^*T_{a+,b-,a,b} & -2ab^*T_{a-,b,a,b-} & -b^*b^*T_{b+,b-,b,b} & M_b^- - \Omega \end{bmatrix} \tag{6.1}$$

where

$$M_a^{\pm} = T_a|a|^2 + 2T_{ab}|b|^2 - 2T_{a\pm,b}|b|^2 - 2T_{a\pm,a}|a|^2 + \Delta_a^{\pm} \tag{6.2a}$$

$$M_b^{\pm} = T_b|b|^2 + 2T_{ab}|a|^2 - 2T_{b\pm,b}|b|^2 - 2T_{b\pm,a}|a|^2 + \Delta_b^{\pm} \tag{6.2b}$$

For a nontrivial solution, when $\det(M) = 0$, it results in a fourth-order equation for Ω, and then, an unstable growth rate for the perturbation is described by Im Ω. For more details on the derivations of these equations, the reader should refer to Gramstad et al. (2018).

This system of equations was used for instability analysis of two crossing Stokes' waves with different crossing angles and wave amplitude and wavenumbers, keeping the same wave steepness for both waves. They found that for small angles of crossing (<67.5 degrees) and for larger angles of crossing (>135 degrees), there is a significant increase in the modulational instability in comparison to the modulational instability of the individual crossing waves (Fig. 6.1). For larger angles, there is an increased growth rate due to the interaction of two waves, while the instability regions of the two original individual waves are slightly decreased. However, for angles close to 90 degrees, there are no or very minimal changes to the modulational instability. These findings were consistent with previous findings by Trulsen et al. (2015) using the CNLS. They also found that a difference in the wavenumber and hence the wavelength of two crossing waves reduces the modulational instability as this difference increases ($s < 1$). The results indicate that interaction between swell and wind seas can increase the modulational instability, providing that the difference in their wavelengths is not too large. The same authors also confirmed these

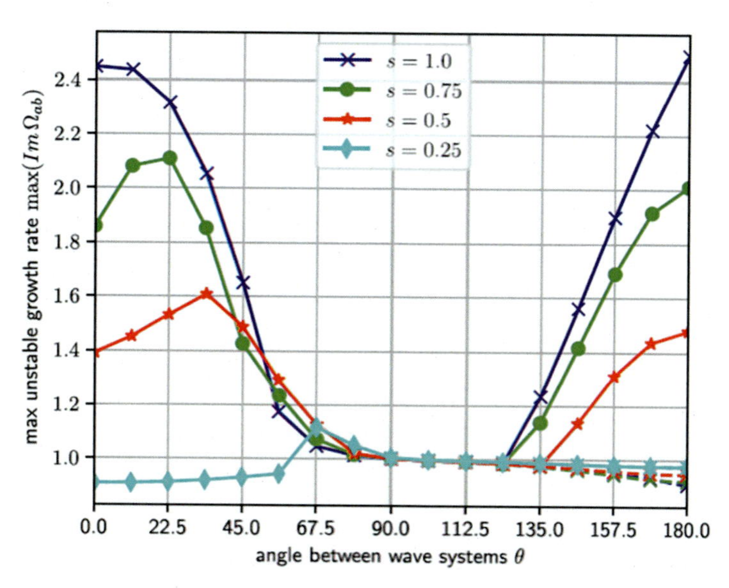

Figure 6.1 The normalized maximum unstable growth rate of the coupled system (max(ImΩ_{ab})/ the maximum growth rate of wave A alone), as a function of angle θ for different values of the wavelength and amplitude ratio s. The dashed lines show the maximum growth rates when instability regions far from the "original" instability regions are excluded. *Source: From Gramstad, O., Bitner-Gregersen, E., Trulsen, K., Borge, J.C.N., 2018. Modulational instability and rogue waves in crossing sea states. Journal of Physical Oceanography 48(6), 1317–1331.*

results through the numerical study of short-crested bimodal wave fields (refer to rogue wave predictions in the crossing wave field section).

Energy-balanced equations for describing crossing seas

Stochastic models for ocean waves have also been developed and used to study mechanisms for the generation of rogue waves in a probabilistic sense. The most common equations used for stochastic modeling are Hasselmann's equation (Hasselmann, 1962), Crawford, Saffman, and Yuen's equation (CSY) (Crawford et al., 1980), and the Alber equation (Alber, 1978). Derivation of the latter two equations is based on the phase-resolved Zakharov and NLS equations, respectively. Randomness is introduced into these equations by assuming the free-wave amplitude spectrum to be a mean zero Gaussian stochastic process. This yields phase-averaged equations, which are used for studying the statistical properties of waves as a result of their evolution and propagation (refer to Stuhlmeier et al., 2019 for the review and further details of these models). The Alber equation is restricted to narrow-banded waves, which is not the case with the CSY equation (Andrade and Stiassnie, 2020b). In fact, Alber's equation can be derived from the CSY equation in the narrowband limit (Andrade and Stiassnie, 2020a).

Recently, these models have been modified with the ultimate aim of studying instabilities and the formation of rogue waves in wind and swell wave systems with different crossing angles. Andrade and Stiassnie (2020a) used the CSY equation to find the instability region of the wave spectrum. Narrow-banded (unstable) JONSWAP spectra were considered in this study. The time evolution was initialized by using bound waves, resulting from the interaction of a stormy sea with a marginal swell, as the initial inhomogeneous disturbance. All the examples considered in this study were found to be unstable to long wave vectors, which implies that not only swell waves but also other long waves such as infragravity waves can trigger the formation of freak waves (Andrade and Stiassnie, 2020a; Rawat et al., 2014). The latter, however, are more relevant for nearshore waves. For broad-banded (stable) spectra, no changes to the nonlinear spectral evolution were observed, indicating that the bound waves arising from the interaction between the local sea and a swell do not generate rogue waves in broadbanded seas. Due to the increasing complexity involved, two-dimensional cases were not studied.

Athanassoulis and Gramstad (2021) extended the Alber equation to describe crossing seas in terms of two quasi-unidirectional wave systems and derived the associated two-dimensional scalar instability condition that controls whether Landau damping or modulational instability is present. Applying these to a selection of wave spectra from the operational spectral wave model provided by the Norwegian Meteorological Institute, they found that only for the most extreme sea states could the spectra be classified as unstable. This agrees with previous studies and indicates that the Penrose–Alber instability is a limiting factor in how narrow a spectrum can be in the ocean. The study found that the findings are sensitive to the selection of the carrier wavenumber. Depending on whether the mean, mode, or median wavenumber was chosen led to different stability/instability classifications for some spectra.

Field and laboratory measurements

Field measurements

Crossing seas have been observed and measured in general sea states (Semedo et al., 2011) as well as in extreme seas during tropical cyclones (Mori, 2012). However, most field measurements of rogue waves are point measurements (Dysthe et al., 2008), and in some cases, the directional distribution is not known. Single-point measurements are normally made either by buoys or radars from offshore platforms (refer to Bitner-Gregersen et al., 2021, for limitations). Measurements like those during a tropical cyclone reported by Collins et al. (2018) demonstrate that bimodal crossing seas form during these events. The measurements of rogue waves were often complemented with numerical simulations to study the evolution of rogue waves, including those in crossing seas (Adcock et al., 2011).

Remote sensing methods such as satellites, X-band radars, and video cameras can cover a much larger area, and the measurements can be used for long- and short-term single-point or spatial statistics. For example, Azevedo et al. (2022) used Sentinel-1 SAR data for identifying rogue waves at the entrance of Tampa Bay, Florida. However, the authors could not identify rogue waves from the background waves in the analysis of the satellite SAR images (Sentinel-1). Apart from these limitations, there is also a lack of ground truth data for the calibration and validation of wave parameters. Nevertheless, the authors identified extreme waves from the wave buoy measurements and found that narrow directional spreading can be one of the indicators of extreme waves.

Stereo video cameras have been deployed at several locations (Fig. 6.2) around the world in the last decade (Fedele et al., 2013; Bergamasco et al., 2017; Benetazzo et al., 2021). They allow for the collection of spatial (order of 100–1000 m^2) and temporal data providing new insight into the formation of rogue waves. Also, crossing seas have been recorded with some of these systems, like during the Kong-rey cyclone (Benetazzo et al., 2021; Davison et al., 2022). There are advantages of these systems, being relatively inexpensive and easily mounted on towers and/or lighthouses or vessels. However, there are also limitations such as lighting, cloud cover, sea spray, viewing angle, and difficulty in deploying in deep water.

Despite these limitations, field measurements have shown that crossing seas could form in different places and when combined with numerical modeling, they contributed to further understanding of crossing sea systems.

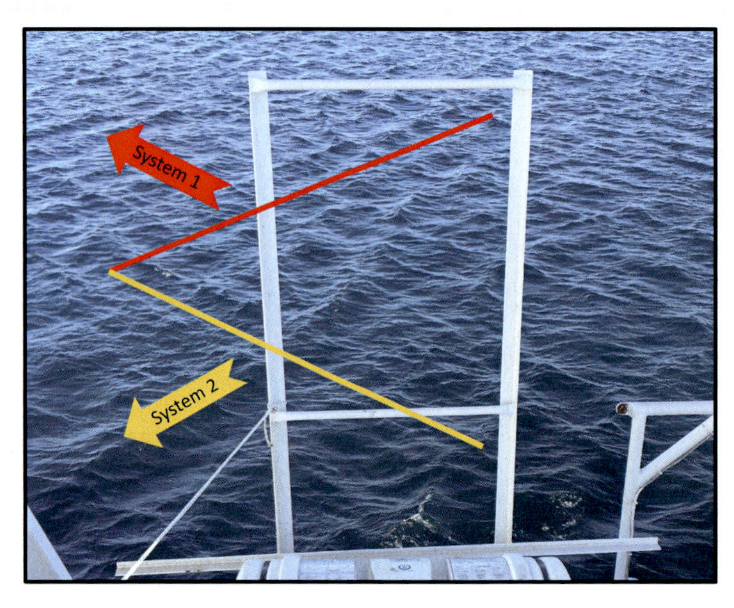

Figure 6.2 Image of a crossing sea state captured by stereo cameras from an oceanographic vessel on March 28, 2017. *Source: Photo credit: Andrea Bergamasco, CNR-ISP; Adopted from Davison, S., Benetazzo, A., Barbariol, F., Ducrozet, G., Yoo, J., Marani, M., 2022. Space-time statistics of extreme ocean waves in crossing sea states. Frontiers in Marine Science 9, 1002806.*

Laboratory experiments

This section summarizes some important experiments involving crossing seas in wave basins. It is important to stress that while the experiments reviewed here used different settings and wave conditions, they all detected rogue waves.

Petrova and Guedes Soares (2009) reported a series of tests in the MARINTEK Ocean Basin facility, which aimed to determine a relationship between the statistical representation of the wave heights and the spectral properties of bimodal unidirectional seas. Three different combined seas were generated, categorized according to the sea-swell energy ratio proposed by Guedes Soares (1984). The data were used to calculate third- and fourth-order moments and the wave height distributions.

Toffoli et al. (2011) examined the effect of the crossing angle on the formation of rogue waves in the same basin in terms of the fourth-order moment of the surface elevation (kurtosis). From the experiment, they found that the kurtosis increased with the angle of crossing up to 40 degrees. As a crossing angle of 40 degrees was the maximum angle that can be generated in the wave basin, tests for the larger angle of crossings were performed numerically.

Experiments by Luxmoore et al. (2019) in the MARINTEK wave basin follow on from Toffoli et al. (2011) and investigate short-crested crossing seas. A range of directional spreading, different angles of crossings, and different separation intervals between peak frequencies were used. In all test cases, rogue waves were detected. However, the kurtosis rarely exceeded the second-order theoretical value. The lowest value of kurtosis was found for crossing sea states with different peak frequencies. Overall, the directional spreading of the individual wave components had more effect on the kurtosis and the wave crest height, and the wave height exceedance probabilities than the angle of crossing.

McAllister et al. (2019) carried out experiments recreating the Draupner wave in the FloWave Ocean Energy Research Facility, University of Edinburgh, UK. Their approach was based on focusing wave groups. They demonstrated that a wave of the same and greater steepness than the Draupner wave can be formed as a result of crossing at large angles (60–120 degrees). They found the presence of a wave set-up forming underneath the highest waves, which was also observed in the field measurements. They also observed that wave breaking limits the maximum achievable wave height in the case of following sea conditions. Further findings from the experiments reported here are summarized in the section on nonlinear wave statistics.

While laboratory studies made important contributions to understanding crossing sea phenomena and the behavior of waves in general, there are laboratory effects and other limitations that can affect experiments. Wave reflection can persist despite methods used to diminish the effect, such as beaches and wave filters in the wavemakers, which in turn can cause the formation of capillary waves. Experiments need to be carefully planned to avoid or reduce the seiching effect. There are also limitations to the size of the crossing angle that can be simulated in wave basins, particularly for multidirectional seas. The measurements are traditionally taken at several points along the main axis of wave basins. However, for multidirectional crossing seas where perturbations can travel along the incident direction of each system and along the main direction of propagation, this configuration of instruments might not capture rogue waves. Also, there is a lack of velocity measurements, which are useful in studying dynamic ocean waves. Recently, new methods, including stereo photogrammetry, were applied to provide better spatial measurements in laboratories (Aubourg et al., 2017; Mozumi et al., 2015).

Rogue wave predictions in crossing wave fields

Numerical simulations

Numerical experiments are often used as complementary to physical experiments but also for investigating phenomena that are difficult to simulate and/

or measure in laboratory experiments and are often used for investigating crossing sea phenomena. The rapid development of computer power and mathematical techniques enabled studies of nonlinear processes over wider computational domains (Xie et al., 2021).

Energy-based equations and modeling of crossing seas

Wave models which solve the energy balance equations and include nonlinear wave interactions such as WAM (ECMWF, 2006) and the third-generation wave model WAVEWATCH III (Tolman, 2014), are used for operational forecasting and can predict bimodal sea states. However, some of the wave models overpredict the dissipation due to white-capping. The MFWAM (Bidlot et al., 2007) with improved physics for dissipation by white-capping has been used for hindcasting sea states with observed rogue waves like the Draupner wave.

The advantage of these models is that they can be applied for forecast and hindcast of wave fields over large spatial (oceans) and temporal scales (days). For example, Tamura et al. (2009) used the WAVEWATCH III model to investigate wave conditions associated with the Suwa-Maru incident. The results indicated that a crossing sea state developed 4 hours before the accident but then became unidirectional at the time of the accident. Interaction between a wind sea and a swell sea developed into a unimodal state with rogue waves. However, these are phase-averaged models, and in studies of crossing seas, they are mainly used in combination with phase-resolving models (Trulsen et al., 2015).

Potential equations and their solutions

Modeling of nonlinear wave phenomena such as rogue waves should ideally be based on fully nonlinear equations (Ruban, 2005). Applying the full nonlinear fluid dynamics equations for larger space- and time-scales and, in particular, for engineering purposes has not been feasible (Fenton, 1999) until recently (Huang et al., 2022). Instead, nonlinear wave modeling using potential flow has been widely applied to predict waves (e.g., around offshore structures). The equations are derived based on assumptions that the fluid is irrotational, inviscid, and incompressible and that the velocity potential must satisfy Laplace's equation. Several numerical and computational methods have been developed to solve these equations (Fenton, 1999) and used for numerical studies of crossing seas. Some of the numerical studies are summarized here; as they are mostly used for deriving statistical parameters of crossing seas, findings are summarized in the section on nonlinear wave statistics.

Wang et al. (2021) applied the Enhanced Spectral Boundary Integral model (Wang and Ma, 2015) to crossing random seas to investigate the impacts of the spectral bandwidth on the changes to extreme wave statistics. The advantage of this computationally efficient method is that it accounts for the overturning of waves as they break. Another powerful approach, developed independently by

Dommermuth and Yue (1987) and West et al. (1987), known as high-order spectral (HOS) method (or HOSM), has been used to solve the free surface Euler equation system. It uses a fast Fourier transform, and computational effort is proportional to the number of wave components N and an arbitrary order of nonlinearity, expressed as a parameter M. Onorato et al. (2006) demonstrated that the method used for computation of the dynamical equations based on the Hamiltonian description of gravity waves can be derived from the HOSM. However, they also noted that these equations are inadequate for studying processes involving wave breaking. Nevertheless, the HOSM has been useful for studies of nonlinear wave–wave interactions and higher-resolution simulations (Brennan et al., 2018). Xiao et al. (2013) included an energy dissipation model to account for wave breaking. HOSM is frequently used for studies of crossing seas, mostly with the order of nonlinearity $M = 3$.

Several authors used HOSM to study the effect of crossing angles on the occurrence of rogue waves. For example, Onorato et al. (2010) used the HOSM for numerical simulations of two nearly unidirectional crossing wave systems to verify the results from the derived CNLS equations. They found that the largest occurrence of rogue waves is found for crossing angles between 40 and 60 degrees and confirmed by an analysis of modulational instability by the same authors. This was further confirmed by Toffoli et al. (2011) using laboratory data and numerical simulation of two long-crested wave systems.

Bitner-Gregersen and Toffoli (2014) expanded the numerical study using HOSM to more realistic directional spread wave spectra to investigate probabilities of rogue wave occurrence. They used WAM hindcast data for several locations around the world oceans (e.g., North Sea) and found that rogue waves can form in wave systems with broad directional spreading. The most recent study, which used HOSM and the energy dissipation formulated by Xiao et al. (2013), investigated the occurrence of rogue waves in crossing seas and their statistical properties (Liu et al., 2022). First, the model was validated with experimental data considering the crossing of two nearly long-crested waves (Sabatino and Serio, 2015; Toffoli et al., 2011) and the crossing of directional seas (Luxmoore et al., 2019). The HOSM results were in good agreement with experimental data of kurtosis and the wave crest height for both cases.

Some numerical studies using HOSM, including those mentioned above, focused on kurtosis. Bitner-Gregersen and Toffoli (2014) found that the kurtosis increases monotonically with the evolution of the wave field. The maximum kurtosis was found for crossing angles between 40 and 60 degrees, as previously by Toffoli et al. (2011) for two long-crested sea states. Gramstad et al. (2018) applied the HOSM along with the nonlinear model based on the Zakharov equations described above. The focus was on the occurrence of rogue waves in short-crested bimodal seas in terms of kurtosis and the maximum observed wave crest. The nonlinear dynamical effects were included in the model by setting $M = 3$ and the wave breaking using the dissipation model by Xiao et al. (2013). It was found that the kurtosis is larger for relatively small and relatively large angles, while minimum kurtosis is found for a crossing angle around 90 degrees. This is in agreement with the findings of Bitner-Gregersen and Toffoli (2014). For the most narrow spectrum used in the simulation ($\gamma = 6$ and $N = 100$), high kurtosis was found for a crossing angle of approximately 40 degrees. This is in agreement with Onorato et al. (2010), who found the maximum kurtosis for crossing angles between 40 and 60 degrees just as the studies mentioned above. They affiliated this with modulational instability, arising from large growth rates and large amplification factors of the instability taking place for these crossing angles. Liu et al. (2022) also investigated the effect of the crossing angles on the kurtosis using the validated HOSM. Two wave systems with JONSWAP spectrum and cosine-squared directional spreading were used for numerical simulations. The wave height and the peak period were the same, while different values for the spectral bandwidth and the directional spreading were used. The crossing angles were 20, 40, 60, and 80 degrees (for more details, refer to Liu et al., 2022). For a wave system consisting of two long-crested crossing waves, the kurtosis increases for the crossing angle between 40 and 60 degrees. For short-crested crossing waves, the kurtosis was independent of the crossing angle.

Moreover, HOSM was used to simulate real sea states and investigate ship accidents. Trulsen et al. (2015) used a direct numerical model to solve the Euler equation shown by Onorato et al. (2006) to be equivalent to the HOSM, to investigate wave conditions surrounding the damage of the tanker Prestige. Hindcast data for the day of the accident were obtained from the wave model MFWAM (ECMWF, 2006). These data showed the development of a bimodal sea with a nearly 90 degrees crossing angle. Davison et al. (2022) used the HOSM, developed by Ducrozet et al. (2016), to simulate several crossing sea states during typhoon Kong-rey in 2018.

Some other modeling studies not described above are by Cavaleri et al. (2012) and Ruban (2009, 2010). Cavaleri et al. (2012) used hindcast from the wave model WAM and CNLS equations to investigate the accident of the cruise ship Louis Majesty on March 3, 2010. It has been suggested that the instability mechanism may be important in explaining the observed wave conditions and the accident. Adcock et al. (2011) used a fully nonlinear quasi arbitrary Lagrangian–Eulerian finite element method potential flow solver to simulate the Draupner wave conditions. Ruban (2009, 2010) used a system of fully nonlinear evolution equations derived for weakly three-dimensional steep water waves (Ruban, 2005) for studying the formation of rogue waves in weakly crossing sea states. It was found that rogue waves have different spatial structures depending on the angle of crossing.

Nonlinear wave statistics

It is apparent from previous sections that one of the main aims of laboratory experiments and numerical modeling is to investigate the probability of rogue wave occurrence and the relationship between kurtosis and rogue wave occurrence. Most of the statistical analysis and statistical models used in the analysis of crossing seas have been used for the analysis of random seas, and hence, it is assumed that the reader will find details elsewhere in this book.

On wave height and wave crest distributions

Early numerical simulations of the wind sea interacting with a plain swell wave, using the Monte Carlo method, showed that the probability of rogue wave occurrence can increase by 5% to 20% in comparison to a corresponding case without swell (Gramstad and Trulsen, 2010). The rogue wave occurrence was found to be dependent on the sea-swell energy ratio by Petrova and Guedes Soares (2010) in laboratory experiments of combined swell and wind waves. The majority of rogue waves occurred in wind-dominated conditions. Bitner-Gregersen and Toffoli (2014) concluded from numerical studies using directionally spread crossing seas that the energy and frequency of the wave systems were the most important factors for occurrence of rogue waves.

Gramstad and Trulsen (2010) also found that rogue wave occurrence depends on the angle of crossing. However, no increase in occurrence in comparison to a corresponding case without swell was found for the angle of the crossing of 90 degrees. Analysis of experimental data of two crossing unidirectional sea states (Toffoli, 2011) showed that the probability of extreme waves increased with increasing crossing angle but was lower than for the unidirectional case (Sabatino and Serio, 2015). Higher wave heights were found for the largest angle of crossing (40 degrees). However, there was no clear relationship between the exceedance of the wave height probability and crossing angles in experiments of short-crested crossing seas (Luxmoore et al., 2019).

Luxmoore et al. (2019) found that the wave height exceedance probabilities in short-crested crossing seas decrease down the tank and could be described by a Forristall distribution (Forristall, 1978) and a Rayleigh distribution at the distance $x/L = 3.2$ and $x/L = 22.4$ from the wavemaker, respectively. Sabatino and Serio (2015), on the other hand, found higher wave heights at the distance $x/L > 15$ from the wavemaker, in an unidirectional crossing sea. Accordingly, wave height distributions deviated from the Rayleigh and Socquet-Juglard second-order distribution (Socquet-Juglard et al., 2005) at the distance $x/L > 22$, which was also mirrored in an increase of the kurtosis. Only for a crossing angle of 10 degrees did the wave heights not exceed those predicted by the Rayleigh distribution. They also calculated the correlation coefficient between successive wave heights. The correlation coefficient was highest for crossing angles of 40 degrees and exceeded the correlation coefficient for unidirectional seas for crossing angles >20 degrees. Petrova and Guedes Soares (2010, 2011, 2014) applied a range of wave height distribution models to the unidirectional swell–wind wave systems.

Similar patterns were observed with the wave crests and their distributions. Just as for the wave heights, Sabatino and Serio (2015) found the highest wave crest for an angle of crossing of 40 degrees and at a distance of $x/L > 15$ from the wavemaker. Similarly, the distribution of wave crests and troughs for bimodal following sea showed dependence on the crossing angle and the propagated distance from the wavemaker (Petrova and Guedes Soares, 2014). However, the crest height was found to be nearly independent of the crossing angle in numerical experiments by Gramstad et al. (2018) of broadband crossing seas. There were few differences in exceedance probabilities of wave crests for cases with different crossing angles in experiments with short-crested crossing seas by Luxmoore et al. (2019).

While the distribution of wave crests deviated from the Rayleigh and Socquet-Juglard second-order distribution in long-crested crossing seas (Sabatino and Serio, 2015), the wave crest height exceedance was grouped around the second-order Tayfun (1980) and Forristall (2000) distributions close to the wavemakers in short-crested crossing seas (Luxmoore et al., 2019). The wave crest height exceedance was underpredicted by the Tayfun distributions further away from the wavemaker. Liu et al. (2022) found that the tail of the exceedance probability of the wave crest amplitude is significantly higher than the second-order (Tayfun, 1980) distribution results for long-crested crossing seas, while they are in good agreement for short-crested crossing seas.

On skewness of surface elevation

A vertical asymmetry of the surface elevation resulting from the sharpening of wave crests and the flattening of wave troughs can be measured by means of the skewness. The skewness describes the importance of the second-order bound nonlinearities (Fedele et al., 2016). Petrova and Guedes Soares (2009) found that the value of skewness is smaller for swell-dominated than for wind-dominated seas. Several studies found that the skewness is dependent on the angle of crossing between two wave components. For example, the skewness increased as the crossing angle decreased toward 22.5 degrees, in a numerical simulation using directional spectra from the Draupner wave by Brennan et al. (2018). Toffoli et al. (2006) simulated unimodal and bimodal directional spectra. They defined a nondimensional parameter, a ratio of the skewness of a directional sea to the skewness of a unidirectional sea, to account for the effect of directional spreading on vertical asymmetry, and found that the parameter decreased with increasing crossing angles. The largest value of the parameter was found for the smallest crossing angle (35 degrees) and the smallest value for the largest crossing angle

(90 degrees). The parameter increased with an increase in the non-linear parameter, which is defined as a function of the wave steepness and the relative depth.

On kurtosis of surface elevation

A strong link between increasing kurtosis and increasing freak wave occurrence was established by Mori and Janssen (2006), and hence, it has been used in the studies of crossing seas as a proxy for the occurrence of rogue waves. The excess kurtosis, which is the difference between the estimated kurtosis and the value of 3, includes a dynamic component due to nonlinear wave–wave interactions (Janssen, 2003) and a bound contribution induced by the characteristic crest–-trough asymmetry of ocean waves. Petrova and Guedes Soares (2009, 2011) found that for wind sea-dominated conditions, in a combined swell–wind sea state, the kurtosis was higher than for swell-dominated conditions.

As reported above, Toffoli et al. (2011) found that the kurtosis increases with increasing angle of crossing up to the maximum value for crossing between 40 and 60 degrees (Fig. 6.3). For angles of crossing larger than 60°, the kurtosis decreased again. The kurtosis increased almost monotonically along the basin for about 15–20 wavelengths and then decreased again (Sabatino and Serio, 2015). Similarly, the kurtosis was found to increase as the crossing angle decreased toward 22.5 degrees in the time series simulated in numerical experiments of crossing seas, potentially contributing to the formation of the Draupner wave (Brennan et al., 2018). A more recent numerical study reported by Gramstad et al. (2018) found a peak in kurtosis at large and small crossing angles for broadbanded crossing seas. A positive correlation between kurtosis and unstable growth was found for two narrow directional spread cases ($N = 16$ and $N = 100$), while there was no correlation for the most broadbanded case ($N = 4$).

Unlike these studies, Luxmoore et al. (2019) did not find any relationship between the kurtosis and the crossing angle in their experiments of crossing

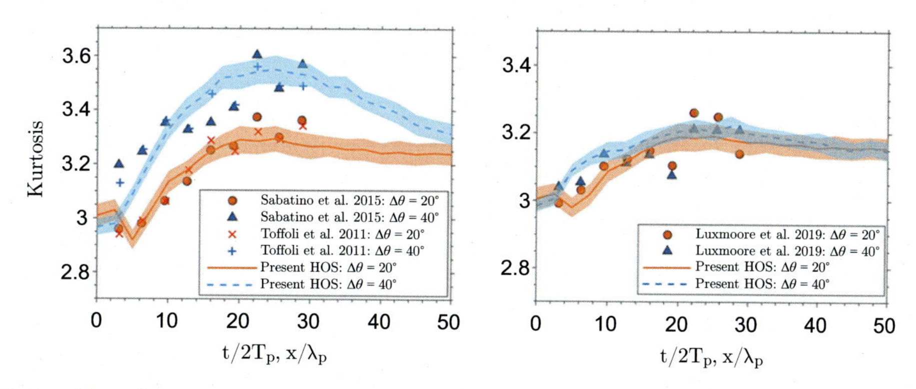

Figure 6.3 Spatial evolution of kurtosis for different angles. *Source: From Liu, S., Waseda, T., Yao, J., Zhang, X. 2022. Statistical properties of surface gravity waves and freak wave occurrence in crossing sea states. Physical Review Fluids 7(7), 074805.*

short-crested waves (Fig. 6.3). The directional spreading of each wave component seemed to have more effect on kurtosis. Liu et al. (2022) established a clear difference between long-crested and short-crested crossing waves. For the former, the kurtosis increases with the crossing angle and reaches the maximum for angles between 40 and 60 degrees. There is no clear relationship between the kurtosis and the crossing angle for the latter.

Luxmoore et al. (2019) validated the Mori et al. (2011) formulae for the kurtosis based on the two-dimensional BFI, which in turn depends on the directional spreading. Generally, the kurtosis estimated, including directional spreading, matched the data. Liu et al. (2022) numerical study confirmed that the kurtosis can be generally estimated by the overall mean directional spreading. However, in the case of crossing long-crested waves, the formulae underpredict the kurtosis. The authors derived a coupled BFI (CBFI) for crossing seas, which includes third-order nonlinearity following Onorato et al. (2010). They showed that the formula for the kurtosis, which included the CBFI, could predict kurtosis reasonably well for a wide range of frequency bandwidths and directional spreadings.

Another numerical study by Wang et al. (2021) investigated the relationship between kurtosis and directional spreading in crossing seas. They found that the kurtosis will become smaller when the spectral bandwidth of each individual random wave component in crossing seas increases. In addition, Luxmoore et al. (2019) found that the kurtosis decreased with increasing separation between peak frequencies of the wave components in crossing seas.

Perspectives

Considerable understanding has been developed in recent years of the conditions leading to rogue waves in crossing seas. Evidence of rogue wave formation in crossing seas came from field, laboratory, and numerical studies. Early studies focused primarily on unidirectional crossing seas, allowing the application of well-known equations to study the effect of angular and frequency separation on rogue wave formation and the modulational instability of such bimodal systems. Recent focus has shifted to more realistic ocean conditions of broadbanded crossing seas, confirming that second-order nonlinearities are responsible mechanisms for the formation of extreme and rogue waves. Most studies focused on the effects of the crossing angle between the two wave components on kurtosis as a proxy for the presence of rogue waves. A clear relationship between the crossing angle and the kurtosis was found for long-crested crossing seas, with the maximum value of kurtosis for crossing angles between 40 and 60 degrees. However, kurtosis was found to be independent of the crossing angle for the short-crested crossing seas. Laboratory and numerical experiments indicate that the directional and frequency spreading, rather than the crossing angle, may have the most influence on the kurtosis. It was found that kurtosis increases with the narrowing of the directional spreading and frequency spectra. It is worth noticing that most studies considered the same peak frequency and spectral shape for the two wave components. So far, the results indicate that kurtosis decreases with increasing separation between the peak frequencies. However, further studies are needed to confirm these results as well as further investigations of the effects of spectral energy.

The wave height and the wave crest height distributions can generally be well described with the existing statistical distributions. Kurtosis of short-crested crossing seas can be estimated by considering the overall mean directional spreading. For long-crested crossing seas, better estimates can be obtained when the effect of third-order nonlinearities is included. It remains to be seen whether the wave parameters should be calculated for combined sea states or for the constituent parts separately (Støle-Hentschel et al., 2018, 2020). Recent studies, especially numerical ones, have shown that a space-time analysis can be a more appropriate approach for the analysis of extreme waves. It is expected that these analyses will be used more with the increasing availability of remote sensing images (Davison et al., 2022). Recent applications of high-order spectral analyses (Ewans et al., 2021; Aubourg et al., 2017) in investigation of nonlinear wave–wave interactions have shown promising results, but more work is needed to ensure that the methods do not account for trivial combinations.

Until now, most of the studies have been based on numerical modeling, mostly using deterministic HOSM. There is scope to use fully nonlinear numerical modeling based on the Navier–Stokes equations (Huang et al., 2022) and stochastic modeling (Athanassoulis and Gramstad, 2021) in the future. Fewer laboratory experiments have been conducted, as these require large wave basins capable of generating crossing seas. Despite their limitations, such as the maximum crossing angle that can be generated and other laboratory effects, these have made and will continue to make significant contributions to the understanding of rogue wave phenomena. The data from laboratory experiments are also important for the validation of numerical models. Recent applications of remote sensing techniques (Aubourg et al., 2017) for laboratory measurements of the sea surface and the wave velocities will further improve the understanding of the spatial evolution of wave spectra and enable spatio-temporal statistical analysis of sea parameters.

There have so far been very few studies based on field data; however, increasing deployment of wave buoys, radar systems, and increasing capabilities of remote sensing will enable more field studies and should enable the testing of theories developed through numerical studies and laboratory experiments.

As with other studies reported elsewhere in this book, past studies focused primarily on steep waves, and wave breaking was not considered. Other effects, such as wind and currents, were not taken into account. With the further development of field and laboratory measurement technologies and increasing computational power, there is scope for further studies of effects not considered so far.

References

Adcock, T.A., Taylor, P.H., 2009. Focusing of unidirectional wave groups on deep water: an approximate nonlinear Schrödinger equation-based model. Proceedings of the Royal Society A: Mathematical, Physical and Engineering Sciences 465 (2110), 3083–3102.

Adcock, T.A., Taylor, P.H., 2014. The physics of anomalous ('rogue') ocean waves. Reports on Progress in Physics 77 (10), 105901.

Adcock, T.A.A., Taylor, P.H., Yan, S., Ma, Q.W., Janssen, P.A.E.M., 2011. Did the Draupner wave occur in a crossing sea? Proceedings of the Royal Society A: Mathematical, Physical and Engineering Sciences 467 (2134), 3004–3021.

Alber, I.E., 1978. The effects of randomness on the stability of two-dimensional surface wave-trains. Proceedings of the Royal Society of London. A. Mathematical and Physical Sciences 363 (1715), 525–546.

Andrade, D., Stiassnie, M., 2020a. Bound-waves due to sea and swell trigger the generation of freak-waves. Journal of Ocean Engineering and Marine Energy 6 (4), 399–414.

Andrade, D., Stiassnie, M., 2020b. New solutions of the CSY equation reveal increases in freak wave occurrence. Wave Motion 97, 102581.

Athanassoulis, A.G., Gramstad, O., 2021. Modelling of ocean waves with the Alber equation: application to non-parametric spectra and generalisation to crossing seas. Fluids 6 (8), 291.

Aubourg, Q., Campagne, A., Peureux, C., Ardhuin, F., Sommeria, J., Viboud, S., et al., 2017. Three-wave and four-wave interactions in gravity wave turbulence. Physical Review Fluids 2 (11), 114802.

Azevedo, L., Meyers, S., Pleskachevsky, A., Pereira, H.P., Luther, M., 2022. Characterizing rogue waves at the entrance of Tampa Bay (Florida, USA). Journal of Marine Science and Engineering 10 (4), 507.

Benetazzo, A., Barbariol, F., Pezzutto, P., Staneva, J., Behrens, A., Davison, S., et al., 2021. Towards a unified framework for extreme sea waves from spectral models: rationale and applications. Ocean Engineering 219, 108263.

Bergamasco, F., Torsello, A., Sclavo, M., Barbariol, F., Benetazzo, A., 2017. WASS: an open-source pipeline for 3D stereo reconstruction of ocean waves. Computers & Geosciences 107, 28–36.

Bidlot, J.R., Janssen, P., Abdalla, S., Hersbach, H., 2007. A revised formulation of ocean wave dissipation and its model impact. In: Technical Memoranda, European Centre for Medium-Range Weather Forecasts (ECMWF), Reading, UK.

Bitner-Gregersen, E.M., Toffoli, A., 2014. Occurrence of rogue sea states and consequences for marine structures. Ocean Dynamics 64, 1457–1468.

Bitner-Gregersen, E.M., Gramstad, O., Magnusson, A.K., Malila, M.P., 2021. Extreme wave events and sampling variability. Ocean Dynamics 71 (1), 81–95.

Brennan, J., Dudley, J.M., Dias, F., 2018. Extreme waves in crossing sea states. International Journal of Ocean and Coastal Engineering 1 (1), 1850001.

Cavaleri, L., Bertotti, L., Torrisi, L., Bitner-Gregersen, E., Serio, M., Onorato, M., 2012. Rogue waves in crossing seas: the Louis Majesty accident. Journal of Geophysical Research: Oceans 117 (C11).

Christou, M., Tromans, P., Vanderschuren, L., Ewans, K., 2009. Second-order crest statistics of realistic sea states. In: Proceedings of the 11th International Workshop on Wave Hindcasting and Forecasting, Halifax, Canada, vol. 18, p. 23.

Collins III, C.O., Potter, H., Lund, B., Tamura, H., Graber, H.C., 2018. Directional wave spectra observed during intense tropical cyclones. Journal of Geophysical Research: Oceans 123 (2), 773–793.

Crawford, D.R., Saffman, P.G., Yuen, H.C., 1980. Evolution of a random inhomogeneous field of nonlinear deep-water gravity waves. Wave Motion 2 (1), 1–16.

Davison, S., Benetazzo, A., Barbariol, F., Ducrozet, G., Yoo, J., Marani, M., 2022. Space-time statistics of extreme ocean waves in crossing sea states. Frontiers in Marine Science 9, 1002806.

De Pinho, U.F., Liu, P.C., Ribeiro, C.E.P., 2004. Freak waves at Campos basin, Brazil. G eofizika 21, 63–67.

Dommermuth, D.G., Yue, D.K., 1987. A high-order spectral method for the study of nonlinear gravity waves. Journal of Fluid Mechanics 184, 267–288.

Ducrozet, G., Bonnefoy, F., Le Touzé, D., Ferrant, P., 2016. HOS-ocean: open-source solver for nonlinear waves in open ocean based on High-Order Spectral method. Computer Physics Communications 203, 245–254.

Dysthe, K., Krogstad, H.E., Müller, P., 2008. Oceanic rogue waves. Annual Review of Fluid Mechanics 40, 287–310.

ECMWF, 2006. IFS Documentation CY36r4. Part VII: ECMWF Wave Model. ECMWF model documentation, technical report.

Ewans, K., Christou, M., Ilic, S., Jonathan, P., 2021. Identifying higher-order interactions in wave time-series. Journal of Offshore Mechanics and Arctic Engineering 143 (2).

Fedele, F., Benetazzo, A., Gallego, G., Shih, P.C., Yezzi, A., Barbariol, F., et al., 2013. Space–time measurements of oceanic sea states. Ocean Modelling 70, 103–115.

Fedele, F., Brennan, J., Ponce de León, S., Dudley, J., Dias, F., 2016. Real world ocean rogue waves explained without the modulational instability. Scientific Reports 6 (1), 1–11.

Fenton, J.D., 1999. Numerical methods for nonlinear waves. Advances in Coastal and Ocean Engineering. World Scientific, pp. 241–324.

Forristall, G.Z., 1978. On the statistical distribution of wave heights in a storm. Journal of Geophysical Research: Oceans 83 (C5), 2353–2358.

Forristall, G.Z., 2000. Wave crest distributions: observations and second-order theory. Journal of Physical Oceanography 30 (8), 1931–1943.

Gibson, R.S., Swan, C., 2007. The evolution of large ocean waves: the role of local and rapid spectral changes. Proceedings of the Royal Society A: Mathematical, Physical and Engineering Sciences 463 (2077), 21–48.

Gramstad, O., Trulsen, K., 2010. Can swell increase the number of freak waves in a wind sea? Journal of Fluid Mechanics 650, 57–79.

Gramstad, O., Trulsen, K., 2011. Fourth-order coupled nonlinear Schrödinger equations for gravity waves on deep water. Physics of Fluids 23 (6), 062102.

Gramstad, O., Bitner-Gregersen, E., Trulsen, K., Borge, J.C.N., 2018. Modulational instability and rogue waves in crossing sea states. Journal of Physical Oceanography 48 (6), 1317–1331.

Grönlund, A., Eliasson, B., Marklund, M., 2009. Evolution of rogue waves in interacting wave systems. EPL (Europhysics Letters) 86 (2), 24001.

Guedes Soares, C., 1984. Representation of double-peaked sea wave spectra. Ocean Engineering 11 (2), 185–207.

Hasselmann, K., 1962. On the nonlinear energy transfer in a gravity-wave spectrum Part 1. General theory. Journal of Fluid Mechanics 12 (4), 481–500.

Huang, L., Li, Y., Benites-Munoz, D., Windt, C.W., Feichtner, A., Tavakoli, S., et al., 2022. A review on the modelling of wave-structure interactions based on OpenFOAM. OpenFOAM® Journal 2, 116–142.

Janssen, P.A., 2003. Nonlinear four-wave interactions and freak waves. Journal of Physical Oceanography 33 (4), 863–884.

Liu, S., Waseda, T., Yao, J., Zhang, X., 2022. Statistical properties of surface gravity waves and freak wave occurrence in crossing sea states. Physical Review Fluids 7 (7), 074805.

Luxmoore, J.F., Ilic, S., Mori, N., 2019. On kurtosis and extreme waves in crossing directional seas: a laboratory experiment. Journal of Fluid Mechanics 876, 792–817.

McAllister, M.L., Adcock, T.A.A., Taylor, P.H., Van Den Bremer, T.S., 2018. The set-down and set-up of directionally spread and crossing surface gravity wave groups. Journal of Fluid Mechanics 835, 131–169.
McAllister, M.L., Draycott, S., Adcock, T.A.A., Taylor, P.H., Van Den Bremer, T.S., 2019. Laboratory recreation of the Draupner wave and the role of breaking in crossing seas. Journal of Fluid Mechanics 860, 767–786.
Mori, N., Janssen, P.A., 2006. On kurtosis and occurrence probability of freak waves. Journal of Physical Oceanography 36 (7), 1471–1483.
Mori, N., 2012. Freak waves under typhoon conditions. Journal of Geophysical Research: Oceans 117 (C11).
Mori, N., Onorato, M., Janssen, P.A., 2011. On the estimation of the kurtosis in directional sea states for freak wave forecasting. Journal of Physical Oceanography 41 (8), 1484–1497.
Mozumi, K., Waseda, T., Chabchoub, A., 2015. 3D stereo imaging of abnormal waves in a wave basin. In: International Conference on Offshore Mechanics and Arctic Engineering. American Society of Mechanical Engineers, vol. 56499, p. V003T02A027.
Okamura, M., 1984. Instabilities of weakly nonlinear standing gravity waves. Journal of the Physical Society of Japan 53 (11), 3788–3796.
Onorato, M., Osborne, A.R., Serio, M., 2006. Modulational instability in crossing sea states: a possible mechanism for the formation of freak waves. Physical Review Letters 96 (1), 014503.
Onorato, M., Proment, D., Toffoli, A., 2010. Freak waves in crossing seas. The European Physical Journal Special Topics 185 (1), 45–55.
Petrova, P.G., Guedes Soares, C., 2009. Probability distributions of wave heights in bimodal seas in an offshore basin. Applied Ocean Research 31 (2), 90–100.
Petrova, P.G., Guedes Soares, C., 2010. Wave height distributions of laboratory generated bimodal seas with abnormal waves. The International Journal of Ocean and Climate Systems 1 (3–4), 239–248.
Petrova, P.G., Guedes Soares, C., 2011. Wave height distributions in bimodal sea states from offshore basins. Ocean Engineering 38 (4), 658–672.
Petrova, P.G., Guedes Soares, C., 2014. Distributions of nonlinear wave amplitudes and heights from laboratory generated following and crossing bimodal seas. Natural Hazards and Earth System Sciences 14 (5), 1207–1222.
Rawat, A., Ardhuin, F., Ballu, V., Crawford, W., Corela, C., Aucan, J., 2014. Infragravity waves across the oceans. Geophysical Research Letters 41 (22), 7957–7963.
Rosenthal, W., Lehner, S., 2008. Rogue waves: results of the MaxWave project. Journal of Offshore Mechanics and Arctic Engineering 130 (2).
Ruban, V.P., 2005. Quasiplanar steep water waves. Physical Review E 71 (5), 055303.
Ruban, V.P., 2009. Two different kinds of rogue waves in weakly crossing sea states. Physical Review E 79 (6), 065304.
Ruban, V.P., 2010. Giant waves in weakly crossing sea states. Journal of Experimental and Theoretical Physics 110, 529–536.
Sabatino, A.D., Serio, M., 2015. Experimental investigation on statistical properties of wave heights and crests in crossing sea conditions. Ocean Dynamics 65, 707–720.
Semedo, A., Sušelj, K., Rutgersson, A., Sterl, A., 2011. A global view on the wind sea and swell climate and variability from ERA-40. Journal of Climate 24 (5), 1461–1479.
Shukla, P.K., Kourakis, I., Eliasson, B., Marklund, M., Stenflo, L., 2006. Instability and evolution of nonlinearly interacting water waves. Physical Review Letters 97 (9), 094501.
Socquet-Juglard, H., Dysthe, K., Trulsen, K., Krogstad, H.E., Liu, J., 2005. Probability distributions of surface gravity waves during spectral changes. Journal of Fluid Mechanics 542, 195–216.
Støle-Hentschel, S., Trulsen, K., Nieto Borge, J.C., Olluri, S., 2020. Extreme wave statistics in combined and partitioned windsea and swell. Water Waves 2 (1), 169–184.
Støle-Hentschel, S., Trulsen, K., Rye, L.B., Raustøl, A., 2018. Extreme wave statistics of counter-propagating, irregular, long-crested sea states. Physics of Fluids 30 (6), 067102.

Stuhlmeier, R., Vrecica, T., Toledo, Y., 2019. Nonlinear wave interaction in coastal and open seas: deterministic and stochastic theory. Nonlinear Water Waves. Birkhäuser, Cham, pp. 151–181.

Tamura, H., Waseda, T., Miyazawa, Y., 2009. Freakish sea state and swell-windsea coupling: numerical study of the Suwa-Maru incident. Geophysical Research Letters 36 (1).

Tayfun, M.A., 1980. Narrow-band nonlinear sea waves. Journal of Geophysical Research: Oceans 85 (C3), 1548–1552.

Toffoli, A., Bitner-Gregersen, E.M., Osborne, A.R., Serio, M., Monbaliu, J., Onorato, M., 2011. Extreme waves in random crossing seas: laboratory experiments and numerical simulations. Geophysical Research Letters 38 (6).

Toffoli, A., Lefevre, J.M., Bitner-Gregersen, E., Monbaliu, J., 2005. Towards the identification of warning criteria: analysis of a ship accident database. Applied Ocean Research 27 (6), 281–291.

Toffoli, A., Onorato, M., Monbaliu, J., 2006. Wave statistics in unimodal and bimodal seas from a second-order model. European Journal of Mechanics-B/Fluids 25 (5), 649–661.

Tolman, H.L., 2014. User manual and system documentation of WAVEWATCH-III version 4.18. Technical Note Nr. 316. NOAA/NWS/NCEP/OMB.

Trulsen, K., Nieto Borge, J.C., Gramstad, O., Aouf, L., Lefèvre, J.M., 2015. Crossing sea state and rogue wave probability during the P restige accident. Journal of Geophysical Research: Oceans 120 (10), 7113–7136.

Walker, D.A., Taylor, P.H., Taylor, R.E., 2004. The shape of large surface waves on the open sea and the Draupner New Year wave. Applied Ocean Research 26 (3–4), 73–83.

Wang, J., Ma, Q.W., 2015. Numerical techniques on improving computational efficiency of spectral boundary integral method. International Journal for Numerical Methods in Engineering 102 (10), 1638–1669.

Wang, J., Ma, Q., Yan, S., Liang, B., 2021. Modeling crossing random seas by fully nonlinear numerical simulations. Frontiers in Physics 9, 593394.

Waseda, T., Kinoshita, T., Cavaleri, L., Toffoli, A., 2015. Third-order resonant wave interactions under the influence of background current fields. Journal of Fluid Mechanics 784 (3), 51–73.

West, B.J., Brueckner, K.A., Janda, R.S., Milder, D.M., Milton, R.L., 1987. A new numerical method for surface hydrodynamics. Journal of Geophysical Research: Oceans 92 (C11), 11803–11824.

Xiao, W., Liu, Y., Wu, G., Yue, D.K., 2013. Rogue wave occurrence and dynamics by direct simulations of nonlinear wave-field evolution. Journal of Fluid Mechanics 720, 357–392.

Xie, J.J., Ma, Y., Dong, G., Perlin, M., 2021. Numerical investigation of third-order resonant interactions between two gravity wave trains in deep water. Physical Review Fluids 6 (1), 014801.

Zakharov, V.E., 1968. Stability of periodic waves of finite amplitude on the surface of a deep fluid. Journal of Applied Mechanics and Technical Physics 9 (2), 190–194.

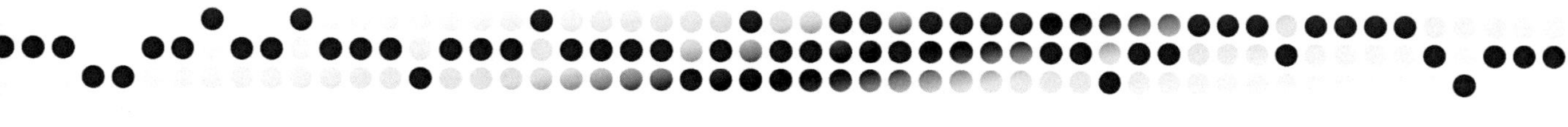

Mechanism 3: Bathymetry effects

Zuorui Lyu
Disaster Prevention Research Institute, Kyoto University, Kyoto, Japan

Introduction

In previous chapters, the definition and some examples of temporal as well as spatial-temporal measurements of rogue/freak waves have been introduced, together with the description of different generation mechanisms to explain the phenomenon. Indeed, in deep-water regions, the contribution from bathymetry to the wave evolution is redundant. As a metaphor, the waves do not "feel" the bottom because the waves' orbital motions decay swiftly with the water depth. However, the bathymetry effects may become important in the occurrence of extreme events in finite and shallow water depth, since the wave deformation occurs due to the significant increase of spatial inhomogeneity.

Based on the dispersion focusing mechanisms, the maximum wave height is related to the four-wave interactions in deep-water. For modulated waves, Benjamin and Feir (1967), Whitham (1974), and Mori and Yasuda (2002) indicated that the effect of the modulational instability on a wave train has a critical depth limit, namely, at a dimensionless water depth of $kh=1.363$. When $kh<1.363$, the modulational instability is not active for unidirectional waves, which implies that the occurrence of the rogue wave at relatively shallow water may significantly decrease due to the attenuation of four-wave interactions. However, when considering ocean rogue wave measurements (Nikolkina and Didenkulova, 2011), it becomes obvious that rogue waves occur not only in deep water but also in finite and shallow water. Research conducted so far indicates that when the waves propagate over an uneven bottom into intermediate and shallow water, the change of the wave depth transforms the dynamics to a nonhomogeneous process, and the variation in high-order cumulants (e.g., kurtosis and skewness) is a key indicator for the occurrence of extreme events.

Science and Engineering of Freak Waves.
DOI: https://doi.org/10.1016/B978-0-323-91736-0.00011-0

Nonlinear wave theory over uneven bottoms

The propagation process of a weakly nonlinear and dispersive wave train can be summarized by a partial differential equation in time and space. Its standard form at a constant depth can be extended to a modified expression with varying depth. In this section, we concentrate on the nonlinear Schrödinger (NLS) equation (Zakharov, 1968; Davey and Stewartson, 1974; etc.), which is the simplest wave evolution equation to account for modulational instability, caused by the interactions from high-order harmonics and the water depth change.

Djordjevic and Redekopp (1978) derived a solution for an envelope-hole soliton moving over an uneven bottom and studied its propagating by means of deriving a depth-modified NLS equation with a slope effect. Variation of the depth in Liu and Dingemans (1989) was divided into different scales, and they gave different types of evolution equations for modulated wave groups over an uneven bottom. The evolution of modulated wave train over an uneven bottom has been also discussed in Peregrine (1983), Turpin et al. (1983), and Mei and Benmoussa (1984).

With the hypothesis of an irrotational, inviscid, and incompressible flow with a free water surface, a coordinate system (x, y, z) is established, as shown in Fig. 7.1. Plane (x, y) is defined along the quiescent water surface, and z is defined in the vertically upward direction, opposite to the gravity acceleration g. The velocity potential ϕ and free surface elevation η are functions of space and time t:

$$\phi = \phi(x, y, z, t), \qquad \eta = \eta(x, y, t). \tag{7.1}$$

In the entire flow field, ϕ is a solution of the Laplace equation to satisfy the continuity equation:

$$\nabla^2\phi = \frac{\partial^2\phi}{\partial x^2} + \frac{\partial^2\phi}{\partial y^2} + \frac{\partial^2\phi}{\partial z^2} = 0. \tag{7.2}$$

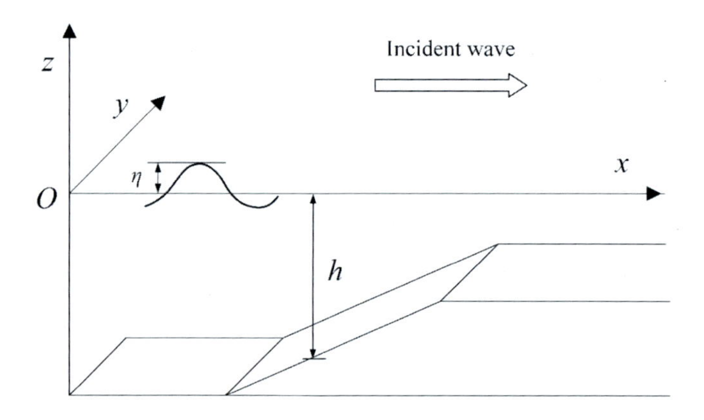

Figure 7.1
Sketch of the wave propagating over an uneven bottom.

On the boundary of free surface $z=\eta(x,y,t)$, ϕ and η satisfy the kinematic boundary condition (i.e., free surface equation) and the dynamic boundary condition (i.e., Bernoulli equation):

$$\frac{\partial\phi}{\partial z}=\frac{\partial\eta}{\partial t}+\frac{\partial\phi}{\partial x}\frac{\partial\eta}{\partial x}+\frac{\partial\phi}{\partial y}\frac{\partial\eta}{\partial y},\quad z=\eta, \tag{7.3}$$

$$2\frac{\partial\phi}{\partial t}+2g\eta+\left(\frac{\partial\phi}{\partial x}\right)^2+\left(\frac{\partial\phi}{\partial y}\right)^2+\left(\frac{\partial\phi}{\partial z}\right)^2=0,\quad z=\eta. \tag{7.4}$$

At the bottom of flow field, ϕ satisfies the no-flux boundary along the sea floor. When the bottom is uneven and water depth varies at $z=-h(x,y)$, ϕ satisfies the equation:

$$\frac{\partial\phi}{\partial z}+\frac{\partial h}{\partial x}\frac{\partial\phi}{\partial x}+\frac{\partial h}{\partial y}\frac{\partial\phi}{\partial y}=0,\quad z=-h(x,y). \tag{7.5}$$

Unidirectional wave propagation

For wave propagating in only one direction (along the x coordinate), the lateral refraction and reflection can be neglected. The water depth h is supposed to be varying slowly. The magnitude of the gradient of depth change satisfies $h'(x)\sim O(\varepsilon^2)$, where ε is the wave steepness. As for the dispersion relation between wave number and frequency ($\omega=\sqrt{gk\lambda}$; $\lambda=\tanh kh$), the carrier angular frequency $\omega=\omega_0$ is constant since there is no temporal variation, but the carrier wave number k changes due to spatial inhomogeneity. Based on the abovementioned inference, we can get $k=k(x)$ and group velocity $c_g=c_g(x)$ on the principal wave direction, and the harmonic term is in the form:

$$E=\exp\left\{\mathrm{i}\left[\int^x k(x)dx-\omega_0 t\right]\right\}. \tag{7.6}$$

These assumptions give a different derivation process of the standard NLS equation with Davey and Stewartson (1974) and lead to the following multiple scales variable substitution for t and x, where the variation of h in the wave evolution process is explicitly reflected:

$$\tau=\varepsilon\left[\int^x\frac{dx}{c_g(\xi)}-t\right],\quad \xi=\varepsilon^2 x. \tag{7.7}$$

We consider the perturbation in two small parameters: the nonlinearity parameter, that is, wave steepness ε, and the modulated parameter δ, which represents the inhomogeneity that comes from the bottom slope and the variation in time and space. With the assumption that ε and δ are of the same order of magnitude, the velocity potential ϕ and free surface elevation η can be expanded into harmonic functions, with small perturbation representing the modulation parameter in a simplified expression only including ε:

$$\phi(x,z,t)=\sum_{n=1}^{\infty}\varepsilon^{n}\left[\sum_{m=-n}^{n}\phi_{nm}(x,z,t)E^{m}\right], \tag{7.8}$$

$$\eta(x,t)=\sum_{n=1}^{\infty}\varepsilon^{n}\left[\sum_{m=-n}^{n}\eta_{nm}(x,t)E^{m}\right], \tag{7.9}$$

The complex conjugate parts satisfy $\phi_{n,-m}=\tilde{\phi}_{nm}$, $\eta_{n,-m}=\tilde{\eta}_{nm}$.

With the method of multiple scales and Taylor expansion, we can give the evolution equation set of wave envelope $A(\xi,\tau)$ in the first harmonic from $O(\varepsilon^{3})E^{0}$ and $O(\varepsilon^{3})E$:

$$\frac{\partial^{2}\phi_{10}}{\partial\tau^{2}}\left(1-\frac{gh}{c_g^{2}}\right)=k^{2}\left[(1-\lambda^{2})+2\frac{c_p}{c_g}\right]\frac{\partial|A|^{2}}{\partial\tau}, \tag{7.10}$$

$$\begin{aligned}
&\mathrm{i}\frac{(1-\lambda^{2})(1-kh\lambda)}{\lambda+kh(1-\lambda^{2})}\frac{d(kh)}{d\xi}A+\mathrm{i}\frac{\partial A}{\partial\xi}\\
&-\frac{1}{2\omega_0 c_g}\left[1-\frac{gh}{c_g^{2}}(1-\lambda^{2})(1-kh\lambda)\right]\frac{\partial^{2}A}{\partial\tau^{2}}\\
&=\frac{k^{4}}{4\omega_0}(9\lambda^{-2}-12+13\lambda^{2}-2\lambda^{4})|A|^{2}A\\
&+\frac{k^{2}}{2\omega_0}\left[2c_p+c_g(1-\lambda^{2})\right]A\frac{\partial\phi_{10}}{\partial\tau},
\end{aligned} \tag{7.11}$$

where phase velocity $c_p=\frac{\omega_0}{k}$. The surface elevation η can be derived from the wave envelope A by multiplication by the harmonic term E. More details on derivation can be found in Djordjevic and Redekopp (1978) and Mei and Benmoussa (1984).

Eq. (7.10) represents the nonlinear frequency correction to the primary wave arising from the mean flow, and it can be added to Eq. (7.11) by introducing the quantity $Q(\xi,\tau)$ following:

$$gQ(\xi,\tau)=\frac{\partial\phi_{10}}{\partial\tau}+\frac{k^{2}c_g}{gh-c_g^{2}}\left(2\frac{c_p}{c_g}+1-\lambda^{2}\right)|A|^{2}. \tag{7.12}$$

Then, the equation set becomes:

$$\frac{\partial Q}{\partial\tau}=0, \tag{7.13}$$

and the equations with respect to wave envelope A are simplified by assuming the solution of Eq. (7.13) is $Q = Q_0(\xi)$:

$$i\beta_h A + i\frac{\partial A}{\partial \xi} + \beta_t \frac{\partial^2 A}{\partial \tau^2} = \beta_n |A|^2 A + \beta_f Q_0 A, \tag{7.14}$$

where

$$\beta_h = \frac{(1-\lambda^2)(1-kh\lambda)}{\lambda + kh(1-\lambda^2)} \frac{d(kh)}{d\xi} = \frac{1}{2c_g}\frac{d(c_g)}{d\xi}, \tag{7.15a}$$

$$\beta_t = -\frac{1}{2\omega_0 c_g}\left[1 - \frac{gh}{c_g^2}(1-\lambda^2)(1-kh\lambda)\right], \tag{7.15b}$$

$$\beta_n = \frac{k^4}{4\omega_0 \lambda^2 c_g}\left\{9 - 10\lambda^2 + 9\lambda^4 - \frac{2\sigma^2 c_g^2}{gh - c_g^2}\left[4\frac{c_p^2}{c_g^2} + 4\frac{c_p}{c_g}(1-\lambda^2) + \frac{gh}{c_g^2}(1-\lambda^2)^2\right]\right\}, \tag{7.15c}$$

$$\beta_f = \frac{k^2}{2\omega_0}\left[2c_p + c_g(1-\lambda^2)\right]. \tag{7.15d}$$

Eq. (7.14) can be further simplified since the term $\beta_f Q_0 A$ only affects the phase of wave. When the $|\tau| \to \infty$ and A tends to zero, Q_0 will also vanish. If one considers a prescribed initial and boundary conditions problem, determining Q_0 is a necessary task. However, if we only concentrate on the evolution process, the value of Q_0 is not important since it only applies a phase shift. Hence, Q_0 can be removed by defining:

$$A(\xi, \tau) = \tilde{A}(\xi, \tau)\exp\left(-i\int^{\xi} \beta_f Q_0 d\xi\right), \tag{7.16}$$

where $\tilde{A}$ is the wave envelope after a phase shift from A. Since we still use A to represent the wave envelope for convenience, Eq. (7.14) can be written as:

$$i\beta_h A + i\frac{\partial A}{\partial \xi} + \beta_t \frac{\partial^2 A}{\partial \tau^2} = \beta_n |A|^2 A. \tag{7.17}$$

Compared with the standard form of NLS equation in Zakharov (1968) and Davey and Stewartson (1974), Eq. (7.17) adds an extra term, reflecting the contribution from spatial inhomogeneity on the dispersion process, and it is proportional to the gradient of water depth change. When $\beta_h = 0$, the modified equation set is equivalent to the standard form for constant water depth from Davey and Stewartson (1974). In Fig. 7.2, we give the variation of β_h, β_t, and β_n in Eq. (7.17) at different dimensionless depths kh. In the deep-water region, the coefficients are independent from kh, which means that a change in the bathymetry does not affect the wave propagation.

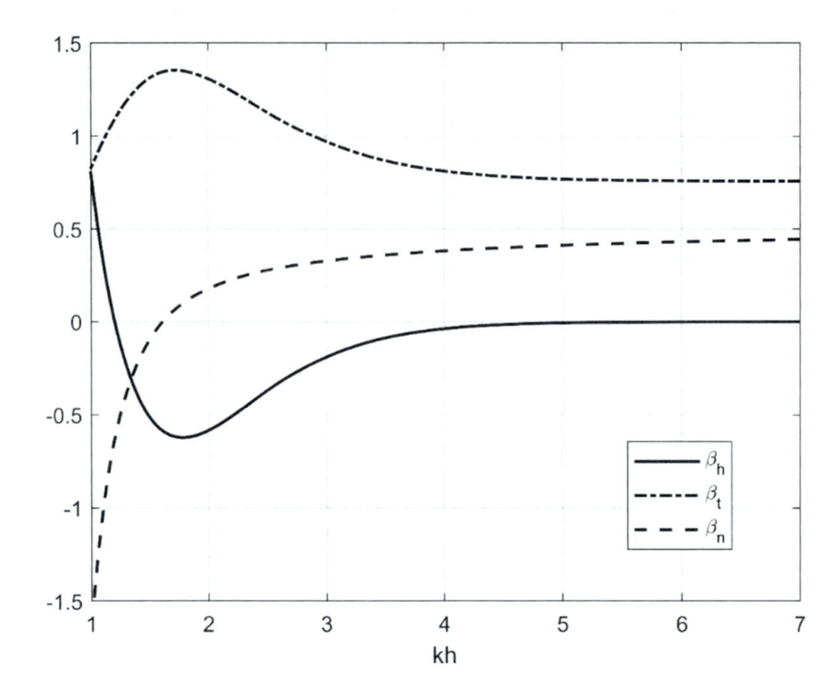

Figure 7.2 Variation of β_h, β_t and β_n at different dimensionless water depths kh when $\omega_0 = 2.51\ s^{-1}$.

A limitation of this theoretical model is that the dispersion effect, due to the depth change on modulated interaction, is restricted to third-order; thus the evolution Eq. (7.17) is only valid for very mild slope cases $h'(x) \sim O(\varepsilon^2)$. A similar equation set can be found in Liu and Dingemans (1989), and the expression of Eq. (7.7)–(7.15c) varies for different definitions of the envelope $A(\xi, \tau)$.

Two-dimensional wave propagation

The theoretical model of modulated wave propagation over uneven bottom in one dimension can be expanded to two dimensions. Liu and Dingemans (1989) put up an evolution model providing more consideration of the wave propagation on the horizontal two dimensions. Assuming that the direction of carrier wave propagation is along the x-axis, the variable substitution into a two-dimensional area is given in:

$$\boldsymbol{x}_1 = \delta \boldsymbol{x} = (\delta x, \delta y), t_1 = \delta t. \tag{7.18}$$

Wave number vector $\boldsymbol{k} = (k_x, k_y)$ and $k = |\boldsymbol{k}|$. Similar to the one-dimension problem, δ and ε are supposed to be of equal order. Then, the velocity potential $\phi = \phi(\boldsymbol{x}, z, t)$ and the free surface elevation $\eta = \eta(\boldsymbol{x}, t)$ can be expanded into a similar perturbation form as in one dimension.

Another expansion in Liu and Dingemans (1989) is that the depth variation can be divided into two parts:

$$h(\boldsymbol{x}) = h_0(\boldsymbol{x}) + \delta h_1(\boldsymbol{x}), \tag{7.19}$$

where $h_0^{'}(\boldsymbol{x}) \approx O(\varepsilon)$, and $h_1(\boldsymbol{x})$ denotes the fast-varying part of depth.

When the carrier wave propagates in the same direction of the gradient of depth change, the evolution equation set for the wave envelope A can be given in the following form:

$$\begin{aligned}
&\frac{\partial A}{\partial t_1} + c_g \frac{\partial A}{\partial \boldsymbol{x}_1} + \frac{1}{2}\left(\frac{\partial}{\partial \boldsymbol{x}_1}\cdot c_g\right)A \\
&\quad - \frac{\mathrm{i}\varepsilon}{2}\left[\frac{\boldsymbol{k}}{k}\frac{\partial}{\partial \boldsymbol{x}_1}\left(\frac{\boldsymbol{k}}{k}\frac{\partial c_g}{\partial k}\cdot\frac{\partial A}{\partial \boldsymbol{x}_1}\right) + \frac{\partial}{\partial \boldsymbol{x}_1}\cdot\left(\frac{c_g}{k}\frac{\partial A}{\partial \boldsymbol{x}_1}\right) - \frac{\boldsymbol{k}}{k}\frac{\partial}{\partial \boldsymbol{x}_1}\left(\frac{c_g}{k}\frac{\boldsymbol{k}}{k}\cdot\frac{\partial}{\partial \boldsymbol{x}_1}\right)A - \mu\frac{\boldsymbol{k}}{k}\frac{\partial A}{\partial \boldsymbol{x}_1}\right] \\
&\quad - \frac{\mathrm{i}\varepsilon}{2} + \mathrm{i}\varepsilon k^2 \omega_0 \kappa |A|^2 A \\
&\quad + \mathrm{i}\varepsilon \frac{\omega_0 k \tanh kh}{2g}(1-\lambda^2)\frac{\partial \phi_{10}}{\partial t_1}A + \mathrm{i}\varepsilon \boldsymbol{k}\frac{\partial \phi_{10}}{\partial \boldsymbol{x}_1}A \\
&\quad + \frac{\mathrm{i}\varepsilon\omega_0}{\sinh 2kh}\left[\left(\mathrm{i}\frac{\boldsymbol{k}}{k}\frac{\partial h_1}{\partial \boldsymbol{x}_1} + kh_1\right) + \frac{\sinh 2kh}{2\omega_0}\nu\right]A = 0,
\end{aligned} \tag{7.20}$$

$$\frac{\partial^2 \phi_{10}}{\partial t_1^2} - \frac{\partial}{\partial \boldsymbol{x}_1}\left(gh_0\frac{\partial \phi_{10}}{\partial \boldsymbol{x}_1}\right) = \frac{g^2}{2\omega_0}\frac{\partial}{\partial \boldsymbol{x}_1}(k|A|^2) - \frac{\omega_0^2}{4\sinh^2 kh}\frac{\partial |A|^2}{\partial t_1} \tag{7.21}$$

where μ, ν are functions of the derivatives of depth and wave number and given in Liu and Dingemans (1989), and

$$\kappa = \frac{1}{16}(9 - 10\lambda^2 + 9\lambda^4) - \frac{1}{2\sinh^2 2kh}. \tag{7.22}$$

The envelope A here satisfies

$$\phi_{11} = -\mathrm{i}\frac{gA\cosh k(z+h)}{2\omega_0 \cosh kh}, \eta_{11} = \frac{1}{2}A. \tag{7.23}$$

We refer to Liu and Dingemans (1989) for more details about the derivation of Eqs. (7.20) and (7.21).

If the slope is very mild and h_0 only varies in the carrier wave direction and the carrier wave rays are parallel straight lines, then, Eqs. (7.20) and (7.21) can be significantly simplified. Along the carrier wave direction, a one-dimensional wave propagation model can be derived. When the fast-varying component $h_1(\boldsymbol{x})$ is ignored, this result is the same as in Djordjevic and Redekopp (1978)'s work.

Modeling and solutions

For the NLS equation in the standard form, we can give some particular solutions as the theoretical formulations such as the decaying solutions, as well as solitons and breathers (e.g., Benney and Newell, 1967; Segur and Ablowitz, 1976; Hui and Hamilton, 1979; etc.). When it comes to an uneven depth, the theoretical solutions for both one dimension and two dimensions are very limited. It should be noted that due to the complexity of the two-dimensional problem, the following discussion will only concentrate on the one-dimensional case. Djordjevic and Redekopp (1978) gave some special solutions for an envelope soliton and an envelope-hole soliton for Eq. (7.14), which has a similar form as the fission process of the Korteweg–de Vries (KdV) equation.

A more general and comprehensive way to discuss the bathymetry effects on the nonlinear modulated wave propagation is by numerically solving the evolution equation like Eq. (7.14) and constructing the wave surface elevation on sufficiently long time and space scales, which is not only limited to the NLS-like equation for an uneven bottom. Several numerical studies of the different nonlinear evolution equations have been given for NLS-like equations (Zeng and Trulsen, 2012; Lyu et al., 2021a; Kimmoun et al., 2021), the KdV equation (Sergeeva et al., 2011; Majda et al., 2019), Boussinesq equations (Gramstad et al., 2013; Kashima et al., 2014; Zhang et al., 2019), and other nonlinear methods (e.g., Viotti and Dias, 2014; Zheng et al., 2020).

Here, we briefly introduce the numerical process of the NLS-like equation, as reported by Zeng and Trulsen (2012) and Lyu et al. (2021a). In the derivation of the NLS-like evolution equation such as Eq. (7.14), the simulation of the modulated wave train is affected by the wave phase and the long wave term Q_0. In other words, the bathymetry effect problem requires the consideration of both, boundary value, and initial condition. If we concentrate on the explicit effect from bottom topography, a widely used method consists of applying the Monte Carlo method for irregular waves with random initial phases, which allows us to neglect the Q_0 and simulate the surface elevation as Eq. (7.17).

In the numerical process, the Fourier transform can be applied to make Eq. (7.17) become an ordinary differential equation:

$$\frac{\mathrm{d}F(A)}{\mathrm{d}\xi} = -\mathrm{i}\beta_n\left|F(A)\right|^2F(A) - \mathrm{i}\omega_\tau^2\beta_tF(A) - \beta_hF(A), \tag{7.24}$$

where $F(A)$ is the Fourier amplitude of wave envelope A with respect to time τ. Then, the free surface elevation η is given in the second-order and second-harmonic to construct a nonlinear modulated wave train:

$$\eta(x,t) = \varepsilon E\eta_{11} + \varepsilon^2\left(E\eta_{21} + E^2\eta_{22}\right). \tag{7.25}$$

Different from the ordinal treatment of the spectral wave modeling, we integrate Eq. (7.25) from the offshore to onshore region assuming periodic boundary conditions in time. In Zeng and Trulsen (2012) and Lyu et al. (2021a), the initial data are given as superposition of waves at different frequencies while satisfying a Gaussian distribution (in the evolution process, the nonlinear interaction will lead to an increasing deviation) in time series with random phases. The modulational instability in the initial conditions is estimated by the ratio of the wave nonlinearity and wave spectral bandwidth [i.e., Benjamin–Feir index (BFI); we refer to Chapter 5 for more details].

Discussion

As examples of the aforementioned numerical method, we give some results on the evolution of modulated waves over an uneven bathymetry and discuss the results.

Fig. 7.3 gives the surface elevation η at different locations in the form of time series by Eqs. (7.24) and (7.25), in which the irregular waves propagate from

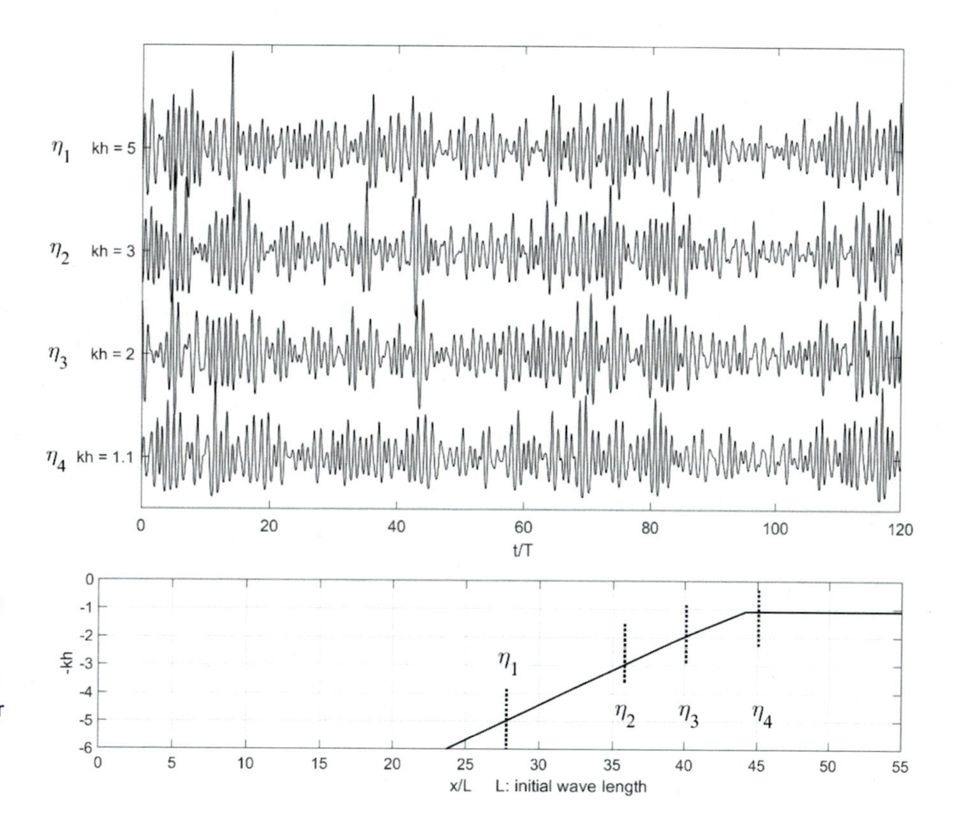

Figure 7.3 Evolution of surface elevation η for a single irregular wave train in time series at a different water depth *kh* over an uneven bottom slope of 0.05.

deep-water to shallow water over an uneven linear bottom with a constant slope equivalent to 0.05; the dimensionless water depth kh changes from 6 to 1.1, with initial BFI $= 0.75$ for $\varepsilon = 0.1$. From Fig. 7.3, it is hard to provide some analysis related to wave height distribution because the occurrence probability of rogue waves is too small while the wave train is irregular. Therefore we apply the Monte Carlo method. We recall that the Monte Carlo method is a mathematical technique based on numerous computational simulations as well as probability and statistics theory. In this study, the Monte Carlo method could help us to investigate the wave height distribution based on the law of large numbers. As the scale of simulations increases, the surface elevation will be close to a strict-sense stationary process, and its statistical properties in the same condition tend to be constant. This is useful to give a quantitative analysis of the differences brought about by a variable, such as different initial wave spectra or topography changes.

By performing large-scale simulations of surface elevation in a wavefield, the occurrence probability of maximum wave height $P(H_{max})$ can be estimated at a reasonable level. For example, $P(H_{max})$ will remain constant at the flat bottom region, and the variation of $P(H_{max})$ over the sloping region can reflect on the contribution from the spatial inhomogeneity. In addition, the fourth-order moment kurtosis μ_4 and the third-order moment skewness μ_3 are also commonly used in the numerical analysis. Referring to Longuet-Higgins (1963), the values of μ_4 and μ_3 in a narrowband Stokes wave model are determined by wave steepness ε. Mori and Yasuda (2002) indicated that μ_4 is related to the quasi-resonant and nonresonant interactions at third-order, which plays an important role in one of the generation mechanisms of rogue waves in deep-water (Chapter 5) and is proportional to the square of BFI. Therefore the kurtosis μ_4 can be applied as an valuable index in the occurrence of rogue waves. As for the skewness μ_3, we can roughly state that it mainly reflects the estimation of nonlinear interaction from a second-order bound wave model.

To practically illustrate this fact with examples, we continue to analyze the Monte Carlo simulation results based on the random wave train from the same initial condition, that is, BFI $= 0.75$ and $\varepsilon = 0.1$, as shown in Fig. 7.3 and Fig. 7.4. We choose two bottom configurations with slopes of 0.05 (also the same as in Fig. 7.3) and 0.01 and then compare the evolution process of the averaged values of μ_4 and μ_3 of surface elevation in the time series and $P(H_{max}), P(\eta_{max})$ as the water depth kh decreases. Based on the standard linear narrow-band wave theory reported by Goda (2000), we take the cumulative probability $P(H_{max} > 8\eta_{rms})$ and $P(\eta_{max} > 4\eta_{rms})$ (they are equivalent in this theory) to compute the occurrence probability of rogue waves. μ_4, $P(H_{max} > 8\eta_{rms})$, and $P(\eta_{max} > 4\eta_{rms})$ monotonically decrease from deep to shallow water, and μ_3 increases in the very shallow region. The deviation from different slopes is small at first, but it becomes significant after $kh < 2$. The slope of 0.05 case gives a higher value in each parameter, which indicates that a

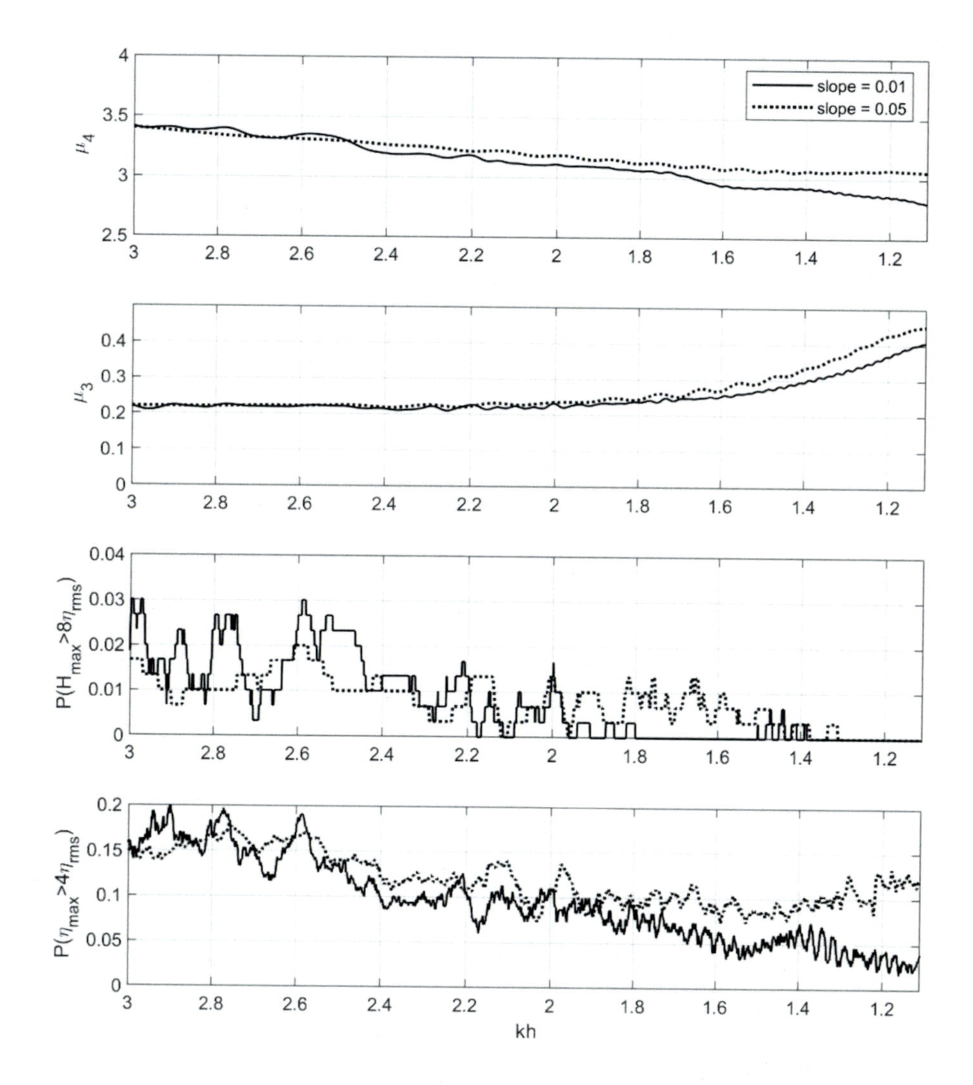

Figure 7.4
Variation of wave statistical parameters at a different water depth *kh* from uneven bottoms with different slopes from BFI = 0.75 and $\varepsilon = 0.1$.

steeper slope will lead to the increase of nonlinear interactions and the occurrence probability of rogue waves. In Fig. 7.4, we only take an ensemble size of 300 in the Monte Carlo simulations; thus the results evaluating $P(H_{max})$ and $P(\eta_{max})$ still have fluctuations around the averaged value, but are sufficient to show the contribution from the bathymetry effect in this example. Estimating rogue waves by directly counting the wave height distribution requires very large ensemble size in Monte Carlo simulation due to the very small occurrence probability. However, the high-order moment parameter is clearer since it points out the evolution of nonlinear effects. In addition, the result between $P(H_{max})$ and $P(\eta_{max})$ has a significant difference in Fig. 7.4, which comes from the increasing second-order contribution in shallow water.

In the numerical simulations performed by Sergeeva, Pelinovsky, and Talipova (2011), Zeng and Trulsen (2012), Gramstad et al. (2013), Viotti and Dias (2014), Majda et al. (2019), Zhang et al. (2019), Lyu et al. (2021a,b), Li et al. (2021), and Kimmoun et al. (2021), a similar conclusion as in Fig. 7.4 can be derived. The increase of the bottom slope angle will give rise to surface instability in very shallow water, which is similar in behavior as when there is increase of occurrence of extreme wave height. The abrupt slope change, like the demarcation point between the sloping section (deep to shallow) and flat bottom, gives a significant increase of $P(H_{max})$ as well as the skewness μ_3, which suggests that the bathymetry effect is more reflected in the second-order effects. If we compare the numerical results in flat bottoms with different water depths, $P(H_{max})$ decreases from the deep to medium water level, while the same applies to the kurtosis μ_4, which corresponds to Benjamin and Feir (1967)'s result on the evolution of modulational instability: in shallower water $kh \leq 1.363$, modulational instability becomes weak or even disappears.

A similar process in shallow water can be verified in laboratory experiments. Physical modeling experiments can give reference for the steeper slope case and more variety of bottom topography, as shown in Trulsen et al. (2012), Kashima et al. (2014), Ma et al. (2014), Whittaker et al. (2017), Bolles et al. (2019), Zhang et al. (2019,2023), Kashima and Mori (2019), Trulsen et al. (2020), and Lawrence et al. (2021). The value of kurtosis μ_4 and skewness μ_3 show a local maximum near the edge between the sloping bottom and flat bottom (deep to shallow). As the slope angle becomes extremely steep, this local maximum reaches its peak at the abrupt depth transitions in shallow water (Zheng et al., 2020; Li et al., 2021). Additionally, the physical experiments give a more accurate estimation of the wave breaking when the wave steepness keeps increasing in shallow water. From Kashima and Mori (2019), the occurrence of wave breaking leads to rapid decrease of kurtosis μ_4 and $P(H_{max})$. Their study also shows that third-order effects are still persistent for steep slopes, which indicates that the wave height in shallow water is not entirely determined by the second-order effect. Mendes et al. (2022) gave a physical explanation about the nonequilibrium wave statistics in wave shoaling, which suggests that the increase of extreme events can be well predicted by the second-order theory.

Based on the numerical simulations and laboratory experiments, the second-order nonlinear effects caused by the bathymetry effects can actually lead to the occurrence of rogue waves when the topography variations are significant. This process can be quantitatively analyzed by systematically adjusting the shape of bottom topography, such as gradually changing the slope angle. It should be pointed out that second-order effects also increase with the increase of wave steepness in very shallow water depth, but it has no contribution to the occurrence of rogue waves if a flat bottom with absence of spatial inhomogeneity is considered.

Summary

In finite water depth, bathymetry effects can lead to the occurrence of rogue waves in particular cases. In this chapter, we briefly introduce the theoretical model of modulated wave evolution over an uneven bottom based on the nonlinear wave interactions, as has been introduced in Chapter 5. The corresponding numerical simulations and experiments provide a consistent conclusion that the second-order effects from bottom topography change play an important role in the occurrence of rogue waves.

There are still many works which will be conducted in the future about bathymetry effects, including but not limited to the following: First, the current theoretical models have a critical limitation assuming that the contribution of the shoaling term is very mild. In other words, a significant bottom topography change, such as a very steep slope, cannot be considered in such an NLS-type wave model. Second, even if we can provide a reasonable propagation model, the required numerical schemes to solve the process is difficult for an uneven bottom in two dimensions. Lyu et al. (2021b) performed simulations of the NLS equation in a two-dimensional directional wavefield; however, the bottom topography is assumed to vary only on the wave principal direction. Additionally, waves in shallow water are subject to breaking before the generation of rogue waves, since they are both the result of strong nonlinear effects. A more precise quantitative analysis considering wave breaking is necessary.

For the development of the study of rogue waves, state-of-the-art research summarized in this chapter discusses another generation mechanism of extreme events, which concentrates on the contribution from bathymetry effects. Based on these results, we can give a more accurate estimation of the wave evolution in offshore and onshore areas with topography information, which provides reference to the extreme wave heights for the navigation and the design of marine structures. This is important to real-world problems, since human activities are more frequent in these areas compared to the deep sea. Generally speaking, a sharp change of the seabed topography will increase the potential risk of rogue waves. From the perspective of scientific exploration, recent studies investigated the evolution of nonlinear waves in an inhomogeneous medium. Their conclusion is sometimes not limited to water waves, but extends also to similar phenomena in other types of fluctuation such as sound or electromagnetic waves. A rapid increase of spatial inhomogeneity, also in these media, may lead to the occurrence of abnormal huge amplitude events during wave evolution.

References

Benjamin, T.B., Feir, J.E., 1967. The disintegration of wave trains on deep water Part 1. Theory. Journal of Fluid Mechanics 27 (3), 417–430.

Benney, D.J., Newell, A.C., 1967. The propagation of nonlinear wave envelopes. Journal of mathematics and Physics 46 (1–4), 133–139.
Bolles, C.T., Speer, K., Moore, M.N.J., 2019. Anomalous wave statistics induced by abrupt depth change. Physical Review Fluids 4 (1), 011801.
Davey, A., Stewartson, K., 1974. On three-dimensional packets of surface waves. Proceedings of the Royal Society of London, A, Mathematical and Physical Sciences 338 (1613), 101–110.
Djordjevic, V.D., Redekopp, L.G., 1978. On the development of packets of surface gravity waves moving over an uneven bottom. Zeitschrift für Angewandte Mathematik und Physik ZAMP 29 (6), 950–962.
Goda Y. 2000. Random Seas and Design of Maritime Structures, 2d ed. *World Scientific*, 464 pp.
Gramstad, O., Zeng, H., Trulsen, K., et al., 2013. Freak waves in weakly nonlinear unidirectional wave trains over a sloping bottom in shallow water. Physics of Fluids 25 (12), 122103.
Hui, W.H., Hamilton, J., 1979. Exact solutions of a three-dimensional nonlinear Schrödinger equation applied to gravity waves. Journal of Fluid Mechanics 93 (1), 117–133.
Kashima, H., Hirayama, K., Mori, N., 2014. Estimation of freak wave occurrence from deep to shallow water regions. Coastal Engineering Proceedings 1 (34), 36.
Kashima, H., Mori, N., 2019. Aftereffect of high-order nonlinearity on extreme wave occurrence from deep to intermediate water. Coastal Engineering 153, 103559.
Kimmoun, O., Hsu, H.C., Hoffmann, N., et al., 2021. Experiments on uni-directional and nonlinear wave group shoaling. Ocean Dynamics 71 (11), 1105–1112.
Lawrence, C., Trulsen, K., Gramstad, O., 2021. Statistical properties of wave kinematics in long-crested irregular waves propagating over non-uniform bathymetry. Physics of Fluids 33 (4), 046601.
Li, Y., Draycott, S., Zheng, Y., et al., 2021. Why rogue waves occur atop abrupt depth transitions. Journal of Fluid Mechanics 919.
Liu, P.L.F., Dingemans, M.W., 1989. Derivation of the third-order evolution equations for weakly nonlinear water waves propagating over uneven bottoms. Wave Motion 11 (1), 41–64.
Longuet-Higgins, M., 1963. The effect of nonlinearities on statistical distributions in the theory of sea waves. Journal of Fluid Mechanics 17 (3), 459–480.
Lyu, Z., Mori, N., Kashima, H., 2021a. Freak wave in high-order weakly nonlinear wave evolution with bottom topography change. Coastal Engineering 167, 103918.
Lyu, Z., Mori, N., Kashima, H., 2021b. Evolution of Nonlinear Directional Random Wave Train from Deep to Shallow Water. Journal of Japan Society of Civil Engineers, Ser. B2 (Coastal Engineering) 77 (2), I_7–I_12.
Ma, Y., Dong, G., Ma, X., 2014. Experimental study of statistics of random waves propagating over a bar. Coastal Engineering Proceedings 1 (34), 30.
Majda, A.J., Moore, M.N.J., Qi, D., 2019. Statistical dynamical model to predict extreme events and anomalous features in shallow water waves with abrupt depth change. Proceedings of the National Academy of Sciences 116 (10), 3982–3987.
Mei, C.C., Benmoussa, C., 1984. Long waves induced by short-wave groups over an uneven bottom. Journal of Fluid Mechanics 139, 219–235.
Mendes, S., Scotti, A., Brunetti, M., Kasparian, J., 2022. Non-homogeneous analysis of rogue wave probability evolution over a shoal. Journal of Fluid Mechanics 939, A25.
Mori, N., Yasuda, T., 2002. Effects of high-order nonlinear interactions on unidirectional wave trains. Ocean Engineering 29 (10), 1233–1245.
Nikolkina, I., Didenkulova, I., 2011. Rogue waves in 2006–2010. Natural Hazards and Earth System Sciences 11 (11), 2913–2924.
Peregrine, D.H., 1983. Water waves, nonlinear Schrödinger equations and their solutions. The ANZIAM Journal 25 (1), 16–43.

Sergeeva, A., Pelinovsky, E., Talipova, T., 2011. Nonlinear random wave field in shallow water: variable Korteweg-de Vries framework. Natural Hazards and Earth System Sciences 11 (2), 323–330.

Segur, H., Ablowitz, M.J., 1976. Asymptotic solutions and conservation laws for the nonlinear Schrödinger equation. I. Journal of Mathematical Physics 17 (5), 710–713.

Trulsen, K., Zeng, H., Gramstad, O., 2012. Laboratory evidence of freak waves provoked by non-uniform bathymetry. Physics of Fluids 24 (9), 097101.

Trulsen, K., Raustøl, A., Jorde, S., Rye, L., 2020. Extreme wave statistics of long-crested irregular waves over a shoal. Journal of Fluid Mechanics 882, R2.

Turpin, F.-M., Benmoussa, C., Mei, C.C., 1983. Effects of slowly varying depth and current on the evolution of a Stokes wavepacket. Journal of Fluid Mechanics 132, 1–23.

Viotti, C., Dias, F., 2014. Extreme waves induced by strong depth transitions: Fully nonlinear results. Physics of Fluids 26 (5), 051705.

Whitham, G., 1974. Linear and Nonlinear Waves. *John Wiley and Sons*, p. 656.

Whittaker, C.N., Fitzgerald, C.J., Raby, A.C., et al., 2017. Optimisation of focused wave group runup on a plane beach. Coastal Engineering 121, 44–55.

Zakharov, V.E., 1968. Stability of periodic waves of finite amplitude on the surface of a deep fluid. Journal of Applied Mechanics and Technical Physics 9 (2), 190–194.

Zeng, H., Trulsen, K., 2012. Evolution of skewness and kurtosis of weakly nonlinear unidirectional waves over a sloping bottom. Natural Hazards and Earth System Sciences 12 (3), 631.

Zhang, J., Benoit, M., Kimmoun, O., et al., 2019. Statistics of extreme waves in coastal waters: large scale experiments and advanced numerical simulations. Fluids 4 (2), 99.

Zhang, J., Ma, Y., Tan, T., et al., 2023. Enhanced extreme wave statistics of irregular waves due to accelerating following current over a submerged bar. Journal of Fluid Mechanics 954, A50.

Zheng, Y., Lin, Z., Li, Y., et al., 2020. Fully nonlinear simulations of unidirectional extreme waves provoked by strong depth transitions: the effect of slope. Physical Review Fluids 5 (6), 064804.

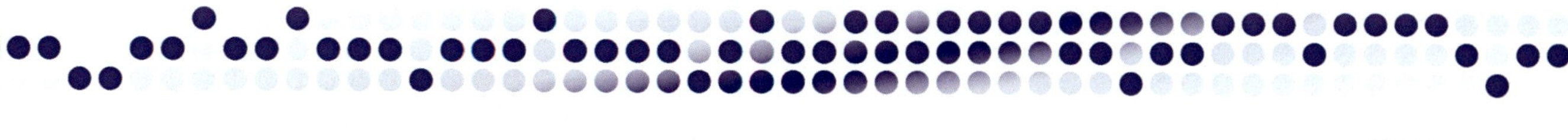

Prediction 1: short-term prediction of extreme waves

Dong-Jiing Doong[1], *Cheng-Han Tsai*[2] *and Chuen-Teyr Terng*[3]
[1]Department of Hydraulic and Ocean Engineering, National Cheng Kung University, Tainan, Taiwan [2]Department of Marine Environmental Informatics, National Taiwan Ocean University, Keelung, Taiwan [3]Marine Meteorology Center, Central Weather Bureau, Taipei, Taiwan

Introduction

Surface gravity waves, which are vehemently random, are some of the most fascinating natural phenomena on the world's oceans. Oceanic activities have risen greatly in recent decades, including sea transportation, the fishing industry, ocean engineering, and recreational activities. The world was estimated to have more than 0.1 million merchant ships and 4 million fishing vessels. Dysthe et al. (2008) reported that more than 22 supercarriers were lost due to severe sea states between 1969 and 1994. Monstrous waves are believed to be the major cause of those shipwrecks, e.g., the Suwa-Maru event (Tamura et al., 2009), Voyager accident (Bertotti and Cavaleri, 2008), Louis Majesty accident (Cavaleri et al., 2012), Onomichi-Maru event (Waseda et al., 2012), and Prestige accident (Trulsen et al., 2015). Marine weather forecasting can help to prevent ship accidents.

Extreme ocean waves are extraordinarily large waves; for example, the significant wave height (H_{m0}) exceeds 10 m. However, waves can be correctly forecasted when the wind-wave generation mechanism is well understood. Freak or rogue waves are defined as waves with individual heights that are at least twice as high as the H_{m0} in the surrounding sea. Freak waves are not necessarily extreme waves since these are isolated events. They appear from nowhere and may impose greater risks on ship navigation than well-predictable extreme waves, especially when the freak waves are also large. Predicting the occurrence of freak waves is challenging because different mechanisms, including linear superposition, directional focusing effects, and nonlinear instabilities, may contribute to their formation; refer to Chapter 1.

Science and Engineering of Freak Waves.
DOI: https://doi.org/10.1016/B978-0-323-91736-0.00008-0

Numerous studies have been conducted on the prediction of freak waves that occur in the open sea. Unlike waves whose generation process has been well formulated, researchers promote the probabilistic forecasting approach as a better way to estimate the possibility of the occurrence of such mysterious and inexplicable freak waves.

However, freak waves appear not only in the deep ocean or shallow water but also on the coast. Coastal freak waves (CFWs) are large amounts of splash water that result from the interaction between shoaling waves and rocks or coastal structures such as breakwaters and armor blocks (Doong et al., 2018). One snapshot of a CFW is shown in Fig. 8.1. Such hazardous waves may sweep people who stand and fish on a breakwater or rock or walk very close to coasts. Tsai et al. (2004) presented that the occurrence of such waves is correlated with the wave grouping effect. Doong et al. (2021) reported that long-period swell increases their occurrence. The generation mechanism of CFWs is more complex than that of freak waves in the open ocean. At minimum, the properties of incident waves, bathymetry, and shoreline categories (rock, beach, or artificial breakwater) influence the occurrence of CFWs. This makes simulating the occurrence of CFWs and determining when and where they will occur very difficult (Chen and Doong, 2022). Since deterministic prediction is difficult, the stochastic method can be an alternative effective approach.

Many freak waves occur due to severe weather systems, such as tropical cyclones, weather fronts, or gale winds. These systems always move and act very quickly. Predicting their statuses for the next few hours is essential. The short-term prediction of freak wave occurrences is sufficient and useful for the safe insurance of ship navigation and coastal activities, which is the main topic of this chapter.

Figure 8.1
A snapshot of a freak wave occurring on the coast.

Prediction with a nonlinear model

The model

According to linear wave theory, the wave height is assumed to follow the Rayleigh distribution. However, wave nonlinearity causes the heights of larger waves to deviate from the Rayleigh distribution. According to Janssen (2003), deviations from the Gaussian distribution of surface elevation can be attributed to the existence of resonant and nonresonant four-wave interactions.

Following Mori and Janssen (2006), the wave height distribution in deep water can be determined by the skewness and kurtosis of the surface elevation (see details in Chapter 5). With assumptions regarding narrow spectrum bandwidth and the independence between the amplitudes and directions of waves, they proposed a maximum wave height distribution and its exceedance probability. Under the linear assumption, the significant wave height is four times the square root of the zero-order momentum. The maximum wave height can be substituted by the definition of a freak wave, i.e., eight times the square root of the zero-order momentum. The probability of a freak wave can be expressed by Eq. (5.29) in Chapter 5. From the equation, the occurrence probability of freak waves is related to the number of waves (N) and the kurtosis of the surface elevation (μ_4). Kurtosis is positively correlated with the freak wave occurrence probability. There must be one freak wave hidden in a very long wave train, e.g., 10,000 waves, which is equal to measuring almost 14 hours of waves.

However, freak waves are highly nonlinear. Mori (2004) suggested that the relation between H_{m0} and total energy m_0 is a quadratic function of kurtosis. The occurrence probability of a freak wave (P_{freak}) is rewritten to Eq. (8.1) with the introduction of variable α, which is equal to 8 under the assumption of linear theory.

$$|P_{freak} = 1 - \exp\left\{ -Ne^{-\left(\frac{\alpha^2}{8}\right)} \left[1 + \frac{\kappa_{40}}{384}\alpha^2\left(\alpha^2 - 16\right)\right] \right\} \tag{8.1}$$

where κ_{40} is a parameter that equals the kurtosis of the surface elevation μ_4 minus 3. α is a parameter that relates to the kurtosis, which can be regressed from field data. Two years (2013/01 to 2014/12) of field data measured by a bottom-mounted ultrasonic wave gauge installed at a water depth of 23 m near Dongji Island in the Taiwan Strait are used. The regression results of α are shown below:

$$\alpha = -0.045\mu_4^2 + 0.406\mu_4 + 6.608 \tag{8.2}$$

It is difficult to validate formula (8.1) with field data because of the limited sampling duration (normally 10 or 20 minutes per hour) for traditional wave stations. However, Dongji makes this validation possible because it is a continuous (60 minutes each hour) sampling station. The verification results are shown in Fig. 8.2. The average nondimensional root mean square error (RMSE) is 13.2%.

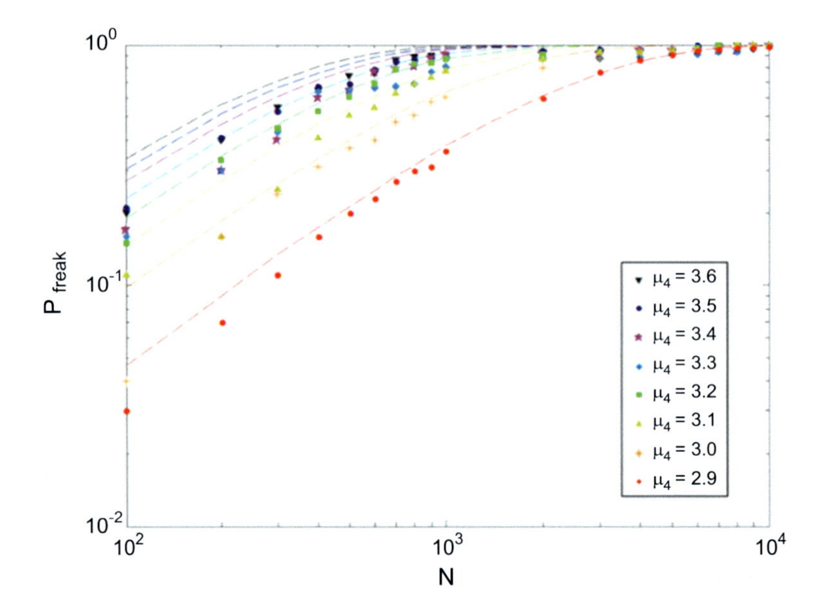

Figure 8.2 Validation of the estimation of occurrence probability of freak waves in the open ocean derived by Eq. (8.1). The validation data are from continuous measurements in the Taiwan Strait.

Operational implementation

The freak wave occurrence probability can be estimated by Eqs. (8.1) and (8.2) when the number of waves N and the kurtosis of the surface elevation μ_4 are known. The number of waves N is related to the mean wave period and the measurement duration. The kurtosis of the surface elevation cannot be directly obtained from the operational wave model. Janssen (2003) suggested that the occurrence of freak waves is due to the modulation phenomenon resulting from the evolution of nonlinear wave effects. The Benjamin–Feir index (BFI), which is the ratio between wave steepness and spectral width, is used to characterize wave instability and is proportional to the root of the kurtosis of the surface elevation. The BFI can be calculated from the wave spectrum. Therefore the spectrum obtained on the grid by the numerical wave model is used to predict the occurrence probability of freak waves. The operational wave model used to construct the prediction model is also implemented operationally. The steps are as follows.

Step 1: Obtain the directional wave spectrum on each grid from the operational wave model.
Step 2: Estimate the BFI considering the wave direction effect. Refer to Mori et al. (2011).
Step 3: Estimate κ_{40} and μ_4 from the BFI.
Step 4: Estimate the occurrence probabilities of freak waves at each grid of the domain according to Eqs. (8.1) and (8.2).

Fig. 8.3 shows a result of a prediction case in which there are barometric highs and lows in the NW Pacific with close intensity. A large wind area is shown and generates rough waves (1 – 1.5 m). The occurrence probabilities of freak waves at the leading edge of the rough wave region are significantly large because of the higher nonlinearity there. Fig. 8.4 shows higher waves and probabilities of freak wave occurrence in the second quadrant of the tropical cyclone movement direction. Fig. 8.5 shows the fast time evolution of the occurrence probabilities over 18 hours. The cross symbol shows a location at which a fishing boat sunk in such a freakish sea state.

The continuous wave observation data of Dongji station are again used to validate the output of the prediction model, as shown in Table 8.1. The assessment classifies events with the same prediction probability from past predictions and

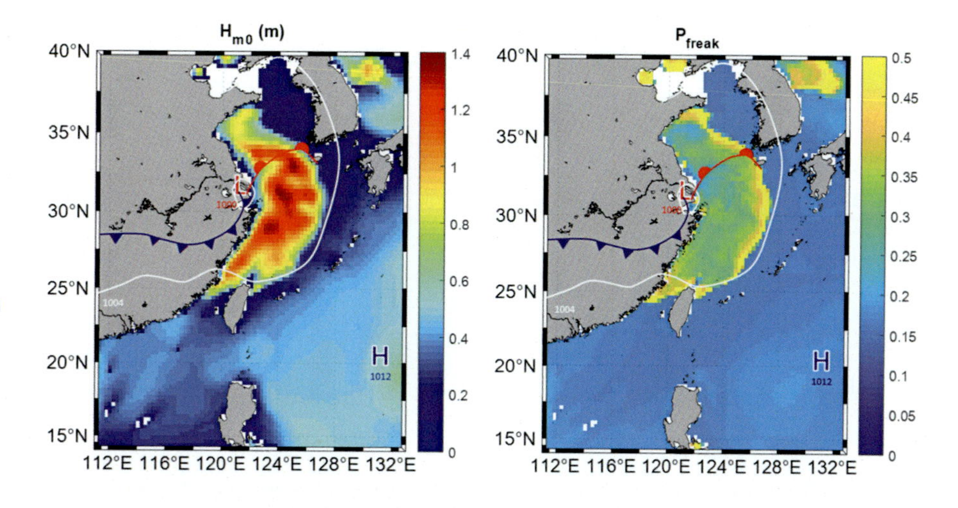

Figure 8.3 Significant wave height distribution (left panel); freak wave occurrence probability (right panel). Time: 2015/9/5 6:00 UTC.

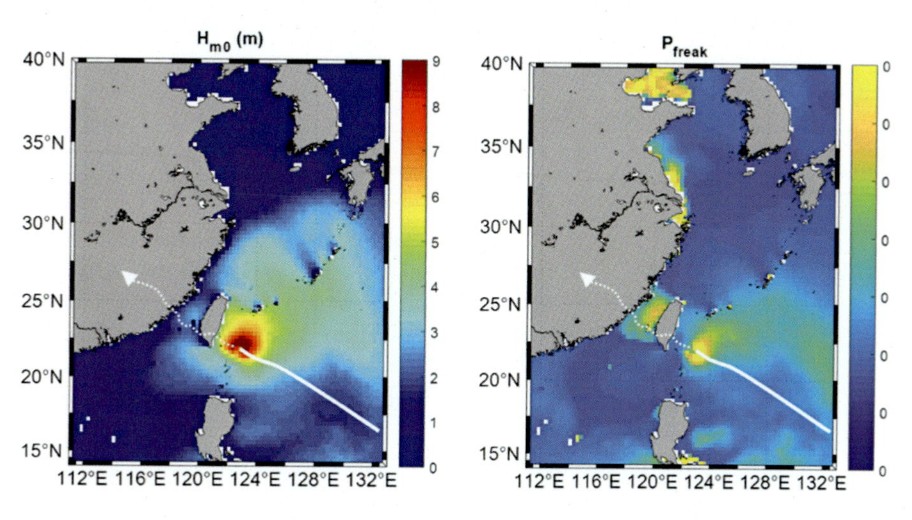

Figure 8.4 Significant wave height (left) and freak wave occurrence probability (right) during typhoon Nepartak (time: 2016/7/7 12:00).

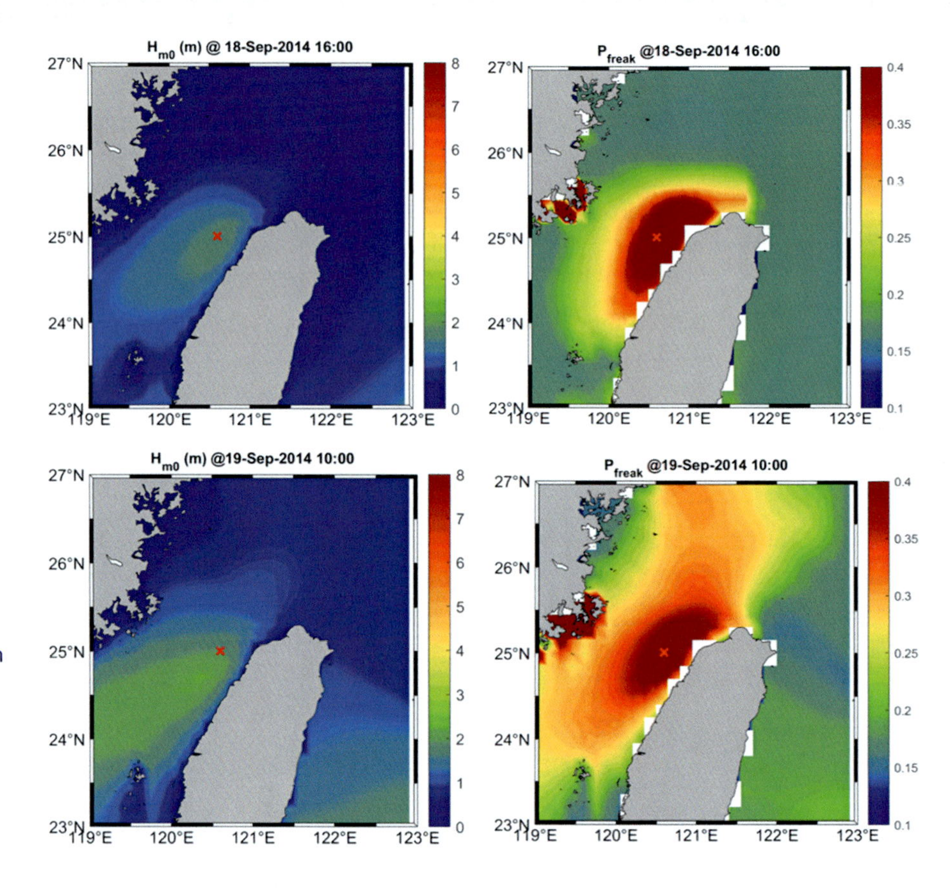

Figure 8.5 Spatial distribution of the significant wave height (left) and P_{freak} (right). The cross symbol points at the shipwreck location.

Table 8.1 Validation of the short-term prediction of freak waves occurring in the open ocean.

Prob. leading time	0.15–0.2	0.2–0.25	0.25–0.3	0.3–0.35	0.35–0.4	0.4–0.45	Weighted average
24 h	14.6%	25.7%	28.5%	27.5%	22.1%	29.4%	15.9%
36 h	13.5%	18.1%	21.6%	22.2%	26.6%	24.4%	22.9%
48 h	18.7%	16.6%	20%	24%	18.3%	22.2%	20.1%

compares it with the actual occurrence probability. The number of samples is used as the weight to calculate the weighted average error. The results show that the predictions with low occurrence probabilities are close to the actual probability with low errors. When the occurrence probability is higher, the validation results also have large deviations. One of the reasons for this is that situations with high probabilities of model operation are relatively rare. The overall average error is between 15% and 22%, which is currently acceptable for such intrinsically random waves.

Prediction with artificial intelligence approaches

Machine learning models

Artificial intelligence, also known as AI, refers to the development of intelligent computers or machines that can exhibit human-like intelligence. AI can be further divided into two methods: machine learning (ML) and deep learning (DL). ML involves analyzing or predicting new events based on previously learned features from data that contain features extracted through human knowledge. On the other hand, DL is based on the theory of neural networks and differs from ML in that it does not require prior knowledge to be extracted by humans. Some of the well-known ML methods include artificial neural network (ANN), support vector machine (SVM), and random forest (RF).

Reikard and Rogers (2011) proposed that although many numerical models have been developed, statistical models are known to be more efficient for short-term forecasting. ML is a data-driven and self-adaptive method for inferring unseen components of problems from data through deep mining, and it can describe nonlinear physical phenomena, which are similar to those of most natural systems. Because of the advantages of AI and the complicated underlying physical mechanisms of CFWs, this method appears appropriate for prediction.

Artificial neural network

ANNs are composed of multiple neurons and are capable of computation and information processing. The architecture of an ANN consists of an input layer, a hidden layer(s), and an output layer. Data are input into the input layer, and the results are output through the output layer, with the hidden layer as the main computation center composed of multiple neurons. Neurons are the basic blocks of ANNs, and information is transmitted from one neuron to another through connections with weights. A single neuron receives inputs from the previous layer's neurons, which are multiplied by the weights and summed with a bias that increases the flexibility of the calculation process, as shown in Eq. (8.3). The result is output through an activation function to control it within a certain range and then output to other neurons.

$$f\left(\sum_{i=1} y_i^l * w_i^{(l+1)} + b_i^{(l+1)}\right) = y_i^{(l+1)} \tag{8.3}$$

where y_i^l denotes the inputs from the previous layer's neurons, $w_i^{(l+1)}$ represents the weights between the previous layer's neurons and the current neuron, $b_i^{(l+1)}$ is the bias added to the computation result which is the output, and f is the activation function. Activation functions include the sigmoid function, hyperbolic tangent function, and rectified linear unit function. The training process of an ANN is based on the gradient descent method that continuously iterates and adjusts the weights in the network to fit the given data. The error gradient is used to find the direction that will reduce the error. The original weights are updated by adding the error gradient multiplied by the learning rate. This process continues until the error is minimized.

Support vector machine

The concept of structural risk minimization is used to find the most suitable data in the given datasets and separate them with a linearly separable function to maximize the margin between the two datasets. An SVM calls the data that determine this function the "support vectors." The hyperplane, which maximizes the margin between the two data groups, is used to determine the best separation of the two groups of data. The goal of an SVM is to find the hyperplane that leads to the farthest distance between the support vectors and the hyperplane, thereby providing the best separation of the two groups of data. For datasets that cannot be separated by a linearly separable function, it projects the data into a higher-dimensional vector space through a kernel function and then finds the hyperplane with the maximum margin between the two groups of data. A hyperplane can be represented as:

$$w^T x + b = 0 \tag{8.4}$$

where w^T is the normal vector of the hyperplane, x denotes the data, and b is a scalar. In an SVM, two boundaries separating the two groups of data are determined. They are represented by two parallel functions from the hyperplane, $w^T x + b = 1$ and $w^T x + b = -1$, where 1 and -1 indicate the classes of the data. The positions of the support vectors are found on the lines parallel to the hyperplane. An SVM finds the maximum value of the distance ($d_{\max}$) between each support vector and the hyperplane. The function that finds the maximum distance is as follows:

$$d_{\max} = \max\left\{\frac{2 * y(w^T x + b)}{w}\right\} = \max\left(\frac{2}{w}\right) \tag{8.5}$$

where $w = \sqrt{w_1^2 + \ldots + w_n^2}$. The function above is derived from the formula for the distance between any point and a line and extended to n-dimensional space. The goal is to solve for the maximum distance from each support vector to the hyperplane. For nonlinearly separable data, an SVM projects the data into a higher-dimensional vector space through a kernel function, including polynomial kernel functions, sigmoid kernel functions, and radial basis kernel functions. The SVM finds the hyperplane that best separates the two given datasets and uses the kernel trick for nonlinearly separable data.

Random forest

An RF involves randomly selecting a data sample to create a decision tree. By repeating this process, multiple decision trees are combined to form a forest. A decision tree consists of several components, including root nodes, nodes, leaf nodes, and branches. The root nodes and nodes are similar in that they use the attributes of the given data to decide how the sample data should be passed down to the next node. There are several paths from the upper node to the lower node, called branches, which represent rules. The relationships between nodes represent the rules for dividing the data and are used to determine the best split points at each node. The data are divided into two groups based on the chosen split rule, and this process continues until all data are completely divided into categories. The leaf nodes represent the results of the categorization procedure.

The rule used to split the data should minimize the sum of squared errors. Given a data sequence $(x_1, y_1), (x_2, y_2), \ldots, (x_i, y_i)$ where x_i denotes the input data in vector form and y_i represents the corresponding class of the input, $i = 1, 2, \ldots, N$, where N is the number of data. If the input data x_i have P attributes $(x_{i1}, x_{i2}, \ldots, x_{iP})$, for a certain attribute j, where $j \in (1, 2, \ldots, P)$, with s as the splitting point, if x_{ij} is larger than s, the data are placed in the R_1 region; if x_{ij} is less than s, they are placed in the R_2 region.

To find the best splitting point s for the data in each node, the sum of squared errors is calculated for the jth input attributes when split at point s. The combination of j and s that results in the minimum sum of squared errors is considered the best splitting point. This is expressed as follows:

$$(j^*, s^*) = \underset{j,\, s}{arg\ min} \left[\min_{c_1} \Sigma_{x_i \in R_1(j,s)} (y_i - c_1)^2 + \min_{c_1} \Sigma_{x_i \in R_2(j,s)} (y_i - c_2)^2 \right] \tag{8.6}$$

where $c_1 = \text{mean}(y_i |, x_i \in R_1)$ and $c_2 = \text{mean}(y_i |, x_i \in R_2)$. This process is repeated until no further split is possible, such as when all the data in a split belong to the same category, and the decision tree is complete. During the tree-building process, a few samples are randomly selected from the dataset. Each decision tree is built by randomly selecting samples and attributes, and the output result is the synthesis of the results of each decision tree.

A comparison among ANN, SVM, and RF is shown in Table 8.2. An ANN has more parameters to adjust, including the number of layers, the number of neurons, the learning rate, and the activation function. The SVM parameters include loss parameters, mapping parameters, and kernel functions, while an RF has fewer parameters, including the number of decision trees and the number of selected sample attributes. ANNs usually take longer to train because

Table 8.2 Comparison between ANN, SVM, and RF.

ML methods	ANN	SVM	RF
Number of parameters	More	Average	Less
Training time	Longer	Shorter	Average
Overfitting problem	Sometimes	Rarely	Seldom
Data quantity	More	Less	Average

they require multiple iterations to adjust their weights and minimize errors, whereas RFs and SVMs take less time to train.

Probability mapping

The original output of an ML method is not a probability value. Mapping from classic results to the probability that we need is the key point. For example, during the ANN building process, an activation function is applied in the hidden layer and the output layer of the network. The tansig transfer function (Eq. 8.7) is generally used to map the output to 0 to 1, which is the range of the occurrence probability.

$$f(\chi) = \frac{2}{1 + e^{-2\chi}} - 1 \tag{8.7}$$

In addition, the original output of an SVM is a binary classification result. To transfer such a result to a probability value, we scale the output to a value between 0 and 1, as shown in Eq. (8.8). This formula is based on the structure of the sigmoid function, with added parameters to fit the output of the model. The probabilities scale the distances from all the data points to the hyperplane.

$$P = \frac{1}{1 + \exp(Af + B)} \tag{8.8}$$

where f is the original output of the model, and A and B are the parameters to be determined, which can be found using maximum likelihood estimation.

An RF is a classification algorithm that uses dichotomy to determine the relationship between the input data and the target. During the training process, an RF creates multiple decision trees and ensembles all the results of each tree. When data are input to the RF model, the model outputs a binary classification, such as 1 or 0. To output the probability result, the average of the outputs from each decision tree is calculated and can be considered the probability of the data output by the RF model. The probability of the RF model can be calculated by the following formula:

$$P = \frac{1}{N}\sum_{n=1}^{N} D_n \tag{8.9}$$

where D_n is the output of each tree, and N is the number of trees. When new data are input into the RF model, the probability result can be obtained based on the average of the results of all decision trees.

Training, validation, and testing

When building a prediction model, it is common to divide the data into three parts: training, validation, and testing. The training data are used to provide the model with the knowledge it needs to learn from, i.e., to fit the parameters in the model. The validation data are used to verify the hyperparameters of the model, such as the number of neurons in an ANN or the number of decision trees in an RF. The testing data are used to evaluate the performance of the model once the training and parameter verification processes have been completed. In the field of ML, it is common to allocate 70% of the given data for training, 10% for validation, and the remaining 20% for testing.

In terms of ML, a confusion matrix can show the results of model predictions in terms of four different outcomes. First, a true positive (A) means that the presence of a condition or characteristic test result is correct. Second, a false positive (B) means that the presence of a condition or characteristic test result is wrong. Third, a false negative (C) means that the absence of a condition or characteristic test result is wrong. Finally, a true negative (D) means that the absence of a condition or characteristic test result is correct.

To assess the accuracy of the constructed model, three indicators are commonly used, including the accuracy rate (ACR), recall rate (RCR), and response rate (RSR). The ACR is an integrated index for assessing the performance achieved on the test, which is defined as the proportion of events correctly predicted by the proposed model. The RCR is defined as the proportion of events that are accurately captured according to the real occurrence events. The RSR is defined as the proportion of events that are accurately captured according to the prediction output by the proposed model. According to the definitions of the indicators, it can be known whether the model overestimates or underestimates results. When the RCR is higher and the ACR is lower, the results show that the model tends to underestimate. Conversely, when the RCR is lower but the RSR is higher, this result shows that the model tends to overestimate the prediction.

Applications

ML methods have recently been implemented to forecast the occurrence of CFWs (Doong et al., 2018, 2020). To build a highly accurate model, as many metocean variables that are correlated with the occurrence of CFWs as possible are included in the model. There are eight inputs to train the ANN model, including the significant wave height, peak wave period, onshore wind speed, misalignment between the wind and wave directions, kurtosis of the sea surface elevation, Benjamin–Feir index, groupiness factor, and abnormality index. The model input data need to be continuously updated in the sense of operational forecasting. Therefore although they are also influential factors such as coastal topography, some variables are not recommended for use due to their low temporal variation or the difficulty of regular access.

Data from 40 actual accidents reported in the news or media are used to build the first ANN model. Fig. 8.6 shows the forecasting results with lead times of 12 and 24 hours, respectively. The abscissa indicates the validation events in series with 23 CFW events (red squares) and 23 non-CFW events (blue circles). Successful forecasts

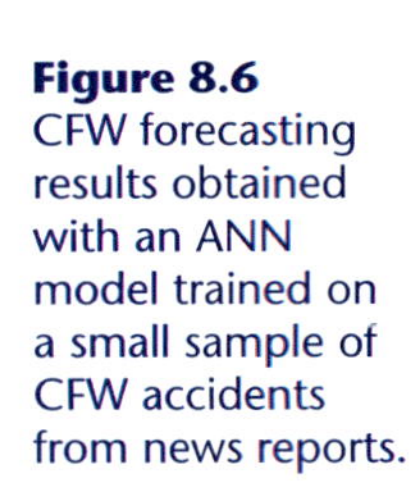

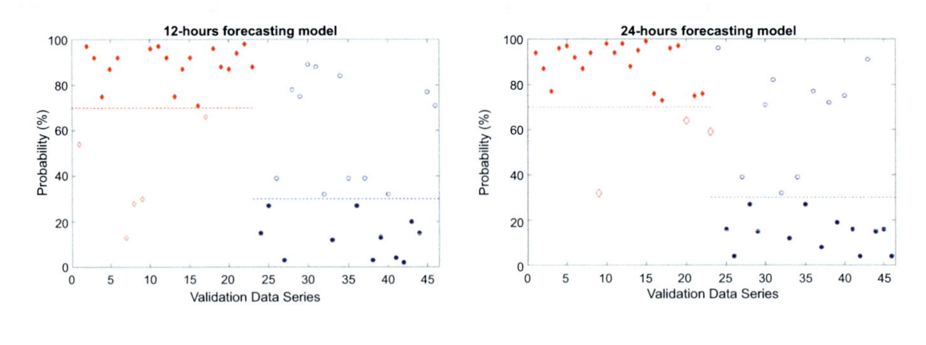

Figure 8.6 CFW forecasting results obtained with an ANN model trained on a small sample of CFW accidents from news reports.

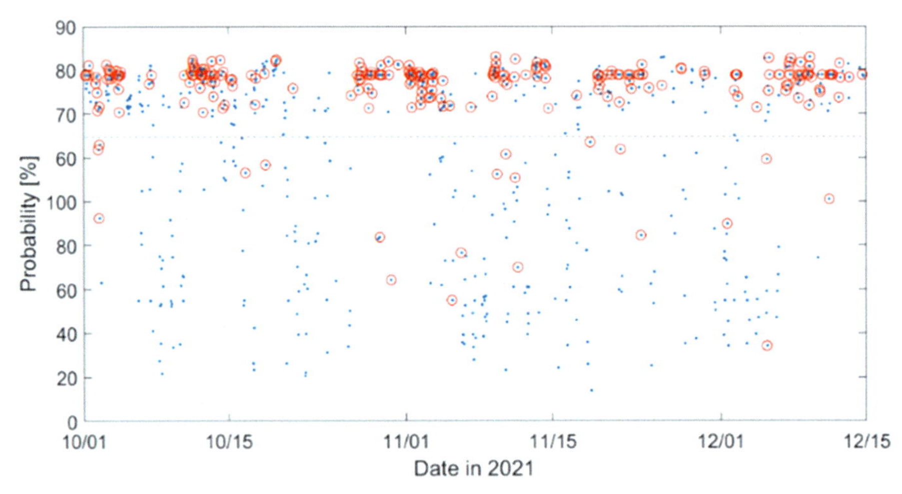

Figure 8.7 CFW forecasting results obtained with an SVM model trained on a large sample of CFW events recorded by a video camera.

are indicated with solid squares and circles, while forecasts that fail are denoted with hollow squares and circles. A probability higher than 70% is assumed to indicate a very high propensity for CFW occurrence, and a probability lower than 30% is regarded as a very low chance of CFW occurrence. The accuracy rate and recall rate for 12-hour forecasting are 63% and 78%, respectively, while those for 24-hour forecasting are 72% and 78%, respectively. The results show that the forecasting model trained with a small CFW accident sample (40 in this article) has a medium forecasting performance level.

The occurrence of CFWs can be recorded by a video camera as shown in Fig. 8.1. The camera monitoring system can catch the CFW occurrences without the misses and biases that normally exist in CFW accidents derived from news reports. An experiment that uses a video camera to monitor CFW occurrences was performed for 75 days continuously, and 600 CFW events were recorded at Longdong on the northeast coast of Taiwan. Fig. 8.7 presents the 12-hour forecasting results obtained by an operational SVM forecasting system. The operational forecasting system continuously updates the forecasts labeled by blue dots in Fig. 8.7. A red circle indicates

Table 8.3 Comparison of the CFW forecasting probabilities (%) obtained with respect to all events and events under severe weather conditions.

	ANN	SVM	RF
Accuracy rate	75.1/78.8	74.4/86.1	71.6/95.1
Recall rate	80.0/98.8	74.0/98.8	73.5/99.4
Response rate	74.3/76.7	74.7/87.1	70.9/92.2

a CFW occurrence recorded by the camera. Generally, CFW events can be caught by the forecasting model with high predicted probability. The accuracy rate and recall rate are 75% and 84%, respectively. The results, compared with those in Fig. 8.6, show that using more and complete CFW data collected from video cameras enhance the achieved CFW forecasting performance with respect to the accuracy rate and recall rate.

The performance of CFW forecasting is found to be connected with the weather conditions and seasons. Table 8.3 lists the CFW forecasting results produced by ANN, SVM, and RF models with respect to 1000 CFW events. The values of the performance are separated by slashes; the numbers to the left and right of each slash indicate the performance results yielded by the forecasting models using all 1000 CFW events and using CFW events observed under severe weather conditions (wind speed >10 m/s), respectively. The comparison results among the three models indicate that different ML models perform similarly with minor differences. However, using all CFW events or using CFW events under severe weather conditions leads to distinguishable performance results. Predicting CFW events under severe weather conditions yields high forecasting performance, implying that sea states or weather conditions play a critical role in forecasting CFWs. CFWs tend to occur under severe weather conditions. In this circumstance, the ML models are better able to model the behaviors of CFW occurrences.

Conclusion

Predicting the occurrence of freak waves is important for sea transportation, the fishing industry, ocean tourism, and engineering design. Unlike ocean wind waves or swells, freak waves are generated by diverse mechanisms, making them very difficult to forecast determinately. This chapter, therefore, introduces two approaches to predict the occurrence probabilities of freak waves in the open sea and coastal areas.

For deep sea freak waves, the formula derived by Mori and Janssen (2006) is revised and used. The formula is validated by field data from a continuous measurement station. Several steps for linking the freak wave prediction model with the operational wave model are proposed. This prediction model is widely examined in various scenarios. The average RMSE of the prediction model varies from 15% to 22%.

For CFW prediction, probabilistic forecasting models have been developed based on ML methods, including ANNs, SVMs, and RFs. Their training processes are introduced. The key challenge in using such a model is to map the original results to probability values. This chapter also assesses the performance of these models. The use of image observations for freak waves is recommended when building a prediction model. Validation results in Longdong of Taiwan show that the performances of the three ML methods are close if the training data are sufficient. The accuracies of the prediction models range from 70% to 80%.

References

Bertotti, L., Cavaleri, L., 2008. The predictability of the "Voyager" accident, Natural Hazards and Earth System Sciences, 8. pp. 533–537.

Cavaleri, L., Bertotti, L., Torrisi, L., Bitner-Gregersen, E., Serio, M., Onorato, M., 2012. Rogue waves in crossing seas: the Louis Majesty accident. Journal of Geophysical Research: Oceans 117.

Chen, Y.C., Doong, D.J., 2022. Modelling coastal freak wave occurrence. Journal of Marine Science and Engineering 10, 323.

Doong, D.J., Chen, S.T., Chen, Y.C., Tsai, C.H., 2020. Operational probabilistic forecasting of coastal freak waves by using an artificial neural network. Journal of Marine Science and Engineering 8, 165.

Doong, D.J., Fan, Y.M., Chen, J.Y., Kao, C.C., 2021. Analysis of long-period hazardous waves in the Taiwan Marine Environment Monitoring Service (TwMEMS). Frontiers in Marine Science 8, 657569.

Doong, D.J., Peng, J.P., Chen, Y.C., 2018. Development of a warning model for coastal freak wave occurrences using an artificial neural network. Ocean Engineering 169, 270–280.

Dysthe, K., Krogstad, H.E., Müller, P., 2008. Oceanic rogue waves. Annual Review of Fluid Mechanics 40, 287–310.

Janssen, P.A., 2003. Nonlinear four-wave interactions and freak waves. Journal of Physical Oceanography 33 (4), 863–884.

Mori, N., 2004. Occurrence probability of a freak wave in a nonlinear wave field. Ocean Engineering 31 (2), 165–175.

Mori, N., Janssen, P.A., 2006. On kurtosis and occurrence probability of freak waves. Journal of Physical Oceanography 36 (7), 1471–1483.

Mori, N., Onorato, M., Janssen, P.A., 2011. On the estimation of the kurtosis in directional sea states for freak wave forecasting. Journal of Physical Oceanography 41 (8), 1484–1497.

Reikard, G., Rogers, W.E., 2011. Forecasting ocean waves: comparing a physics-based model with statistical models. Coastal Engineering 58, 409–416.

Tamura, H., Waseda, T., Miyazawa, Y., 2009. Freakish sea state and swell-windsea coupling: numerical study of the Suwa-Maru incident. Geophysical Research Letters 36, 36280.

Trulsen, K., Nieto Borge, J.C., Gramstad, O., Aouf, L., Lefèvre, J.M., 2015. Crossing sea state and rogue wave probability during the Prestige accident. Journal of Geophysical Research: Oceans 120, 7113–7136.

Tsai, C.H., Su, M.Y., Huang, S.J., 2004. Observations and conditions for occurrence of dangerous coastal waves. Ocean Engineering 31, 745–760.

Waseda, T., Tamura, H., Kinoshita, T., 2012. Freakish sea index and sea states during ship accidents. Journal of Marine Science and Technology 17, 305–314.

Prediction 2: long-term prediction of extreme waves

Francesco Barbariol[1], *Jean-Raymond Bidlot*[2] *and Alvise Benetazzo*[1]
[1]Institute of Marine Sciences (ISMAR), National Research Council (CNR), Venice, Italy
[2]ECMWF, Reading, Berkshire, United Kingdom

Introduction

In this chapter, we focus on the numerical prediction of extreme individual waves over regional-to-global spatial and short-to-long temporal scales. These are the typical scales of numerical model applications, such as marine weather forecasting and climatological studies. Indeed, in addition to the standard wave products—for instance, the significant wave height, mean wave period, and mean wave direction—numerical wave model capabilities allow today to compute the largest individual waves expected in stormy seas. This information is relevant for engineers for designing offshore structures and vessels, for companies and organizations in ship routing—weather routing specifically—and in general for all the actors involved in maritime safety.

Regardless of the spatial and temporal scale of interest, the key question we aim to answer is: what is the largest wave height that one may expect in a given sea state? In this chapter, we show that we can answer that question by relying on phase-averaged numerical wave models, which are the ones routinely used in operational wave forecasting. These models, also called spectral wave models, represent sea waves using the directional energy spectrum. The spectrum is computed at each node of the numerical model grid at each time step, assuming a stationary and homogeneous sea state. Thus, depending on the domain size and temporal coverage, the model produces an ensemble of sea states and directional spectra, and the largest individual waves can be estimated for each of them. How it is achieved will be explained in detail in the next section. It is important to mention that the state-of-the-art spectral wave models are equipped with routines that provide the height of the expected maximum waves as outputs. These

Science and Engineering of Freak Waves.
DOI: https://doi.org/10.1016/B978-0-323-91736-0.00003-1

products do not exclusively describe freak waves, but they represent the extreme waves for a given sea state, whether they exceed the freak wave threshold or not.

Historically, the numerical prediction of the largest waves using phase-averaged wave models began in 2008 with the "freak wave warning system" from the ECWAM model of ECMWF (Chapter 8 in ECMWF, 2021). The system is operational, allowing a global forecast of the expected maximum wave envelope height (details in the next sections). Later, starting from version 5.16, the WAVEWATCH III model (from now on WW3) was developed to provide the expected maximum crest and crest-to-trough heights (Barbariol et al., 2017; WW3DG, 2016; https://github.com/NOAA-EMC/WW3, last visit January 23, 2023). Finally, the MyWave WAM model from cycle 6 was equipped with expected maximum crest and crest-to-trough height outputs (Benetazzo et al., 2021a; https://github.com/mywave/WAM, last visit January 23rd, 2023). In research mode, ECWAM can produce output similar to WW3 and WAM.

These extreme wave model products have been validated in past years against various observational datasets in coastal seas and open oceans. They can now be used for numerical wind–wave prediction, hindcasting, and climatological applications.

Extreme waves from numerical wave models

The basic principle of extreme wave computation within numerical wave models is the combined use of extreme value analysis and wave spectral moments. Indeed, given the intrinsic randomness of the wave process, the prediction of the largest individual wave heights using wave spectra cannot be anything but statistical. Therefore, state-of-the-art probability distributions of maximum wave height are used to estimate, on the one hand, the probability of exceedance of specific thresholds (e.g., the freak wave limit $1.25H_s$ for a crest and $2.0H_s$ for wave heights) and, on the other hand, the most probable (mode) and the expected (mean) value of the variable of interest. As it will be shown later, wave parameters derived from the moments of the directional wave spectrum are used as the parameters of the probability distributions. In other words, the probability of occurrence of extreme waves, their most probable value, and their expected value are expressed as a function of the model directional spectrum $S(k,\theta)$ and its moments:

$$m_{ijl} = \iint k_x^i k_y^j \omega^l S(k,\theta) dk d\theta \tag{9.1}$$

where ω and θ are intrinsic angular wave frequency and wave direction, (k_x,k_y) are the components of the wavenumber vector $\vec{k}$ (with magnitude k) in the east and north directions, respectively, and no contribution from a diagnostic spectral tail after maximum discretized frequency is taken into account for integration.

Theoretical framework

The probability distributions used in spectral wave models to estimate maximum waves are among the most widely employed by scientists and engineers, reflecting the state-of-the-art extreme wave analysis. Two domains of analysis can be considered: one considers the wave maximum occurs at a fixed point at sea during a given time duration Γ; the other considers the wave maximum occurs within Γ and over a given space of fixed sea surface area A (with side lengths X and Y, such that $A = XY$). Accordingly, in the following, we will distinguish between time and space-time extreme crest heights $C_{max,T}$ and $C_{max,ST}$ (with subscript T and ST meaning time and space-time, respectively) and time and space-time extreme crest-to-trough height $H_{max,T}$ and $H_{max,ST}$. At the same time, we only define the time extreme envelope height $E_{max,T}$, as a space-time framework for envelope heights is not presently available. Definitions of time and space-time extremes are given in Figs. 9.1 and 9.2, respectively.

Time extremes

The prediction of the extreme crest and wave heights at a fixed point can rely on several parent probability distributions, from the exceedance distribution functions (EDFs) based on the Rayleigh model and valid for linear and narrow-banded waves, to more sophisticated EDFs accounting for nonlinearities and arbitrary bandwidth effects (Tayfun and Fedele, 2007; Fedele and Tayfun, 2009). Whichever the chosen EDF, it is customary to assume that for a large number of waves in the domain of analysis, the probability distribution of maxima can be well approximated by the Gumbel (1958) distribution, whose expected value is easily evaluated as:

$$z_{max} = \hat{z}_{max} + \frac{\gamma}{\alpha_N} \tag{9.2}$$

where z can be alternatively crest height C, wave height H, or envelope height E, $\gamma = 0.5772$ is the Euler–Mascheroni constant, and $\hat{z}_{max}$ and α_N are, respectively, the mode of z_{max} and the intensity function, depending on the parent probability distribution (Tayfun and Fedele, 2007).

Aimed at providing guidance to marine structure designers for modeling, analysis, and prediction of environmental conditions, some classification societies have issued recommendations on the EDFs to choose for evaluating maximum crest and wave heights. According to one of them, the DNV-GL (2019), the probability $P(C_{max,T} > c)$ that the maximum crest height in a second-order nonlinear sea state with an average number of waves N exceeds a threshold c (e.g., the freak wave crest definition $1.25H_s$) can rely on the Forristall (2000) distribution P_F, such that:

$$P(C_{\max,T} > c) = 1 - (1 - P_F)^N = 1 - \left(1 - \exp\left[-\left(\frac{c}{4\alpha_F\sigma}\right)^{\beta_F}\right]\right)^N \tag{9.3}$$

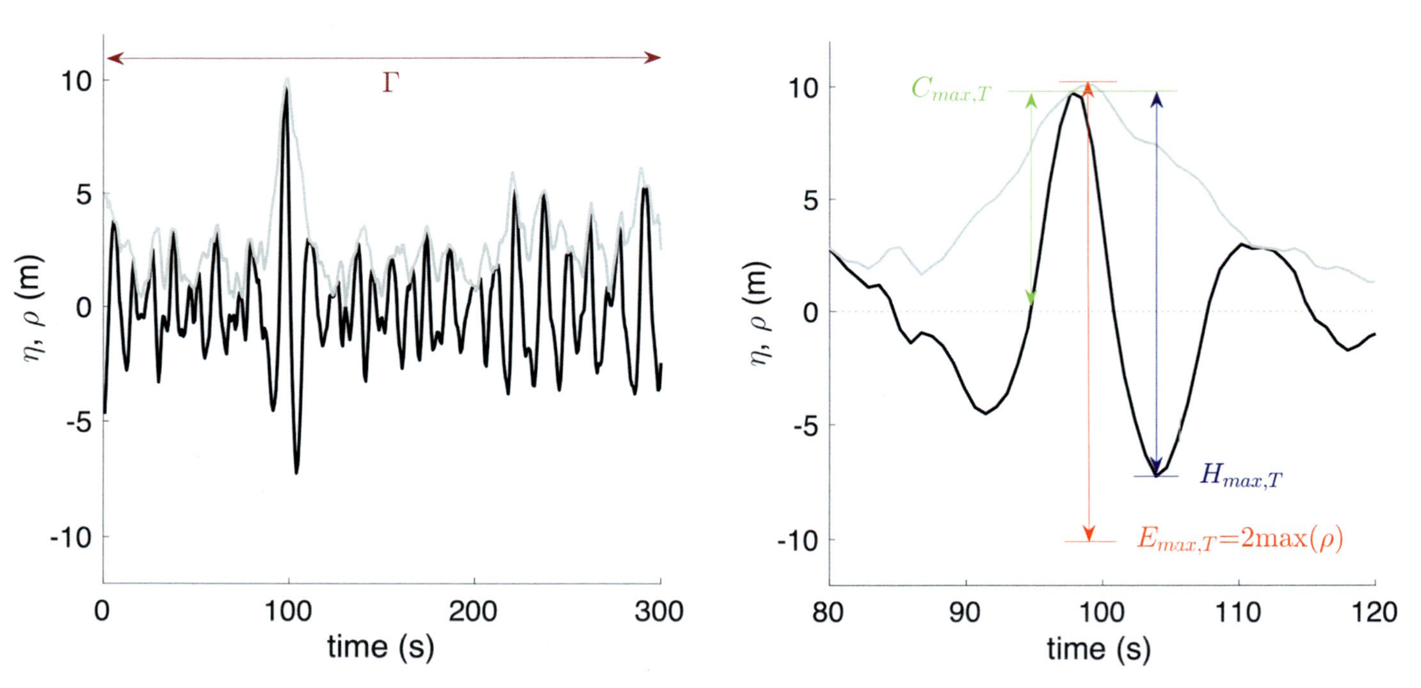

Figure 9.1 Definition of time extremes. Left panel: time series of the sea surface elevation η (solid black line) and its envelope ρ (solid gray line) from the Hilbert transform of the sea surface elevation. The largest wave of the record within duration Γ is visible around 100 s. Right panel: definition of the maximum crest (green), crest-to-trough (blue), and envelope (red) heights of the record. Maximum crest and crest-to-trough heights are defined after zero-crossing analysis of η, while maximum envelope height is defined as twice the maximum of the envelope ρ.

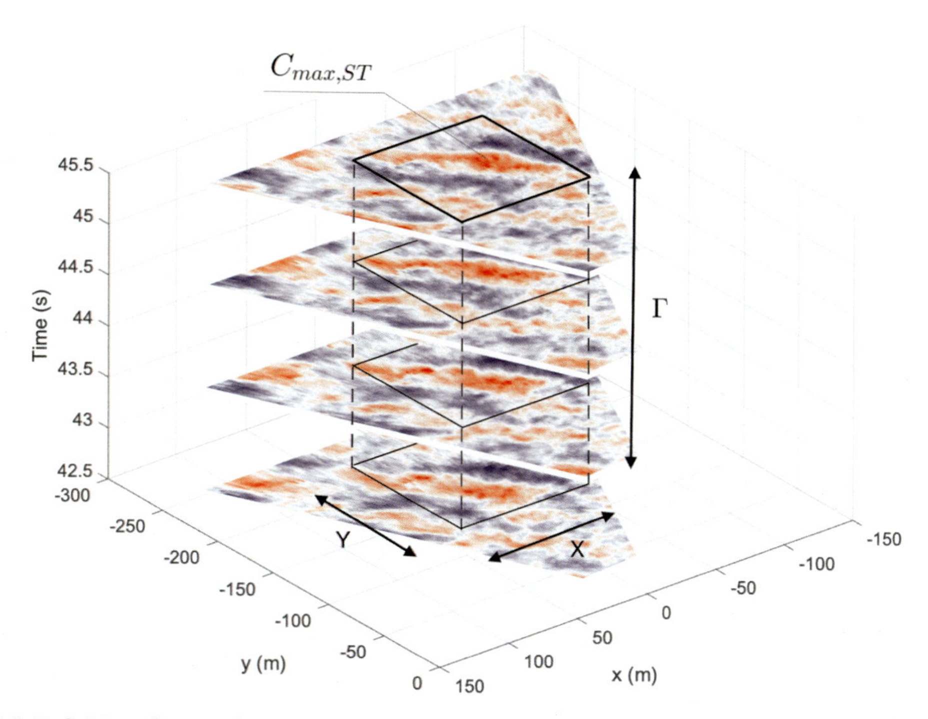

Figure 9.2 Definition of space-time extremes. Maps of the sea surface elevation field measured with a stereo-video imagery system, evolving over space and time (colormap ranges from blue, representing negative values of η, to red, positive values of η). The space-time volume defined by area $A = XY$ and duration Γ is displayed, as well as the largest sea surface elevation within the space-time volume, which can be considered the maximum crest height $C_{max,ST}$ in $A\Gamma$. As the spatial crest-to-trough height is not uniquely defined (it depends on the direction considered), $H_{max,ST}$ is defined as the largest among the zero-crossing crest-to-trough heights obtained from the time series in the space-time domain.

where $\sigma = \sqrt{m_{000}}$ is the standard deviation of sea surface elevation, and α_F and β_F are the parameters of Forristall distribution (depending on nonlinearity and directionality of the sea state; Forristall, 2000). The average number of waves is estimated as $N = \Gamma T_z^{-1}$, being T_z the average zero-crossing wave period, fairly well approximated by the spectral mean period $T_2 = 2\pi\sqrt{m_{000}m_{002}^{-1}}$. After Gumbel approximation (Eq. (9.2)), the expected maximum crest height $\overline{C}_{max,T}$ at a fixed point can be then expressed as:

$$\overline{C}_{max,T} = 4\sigma\alpha_F(\ln(N))^{1/\beta_F}\left(1 + \frac{\gamma}{\beta_F \ln(N)}\right) \tag{9.4}$$

To evaluate the probability $P(H_{max,T} > h)$ that the maximum crest-to-trough height at a fixed point exceeds a threshold h (e.g., the freak wave definition $2H_s$), the

DNV-GL recommends relying on the linear Naess (1985) model P_{Na}, in a way that:

$$P(H_{\max,T} > h) = 1-(1-P_{Na})^N = 1-\left(1-\exp\left[-\left(\frac{1}{4(1-\psi^*)}\left(\frac{h}{\sigma}\right)^2\right]\right)^N \quad (9.5)$$

where ψ^* is the narrow-bandedness parameter, taking into account arbitrary spectral bandwidth effects on the wave statistics. It ranges from 0 to −1, the latter value corresponding to an infinitely narrow-banded sea state, but it typically assumes values around 0.6–0.7 in unimodal wind sea states (Boccotti, 2000). There are different interpretations of how to estimate ψ^*, requiring the evaluation of the autocovariance function $\psi(\tau)$ (Fig. 9.3):

$$\psi(\tau) = \frac{1}{\sigma^2}\iint \cos(\omega\tau)\frac{S(k,\theta)}{c_g}d\theta d\omega \quad (9.6)$$

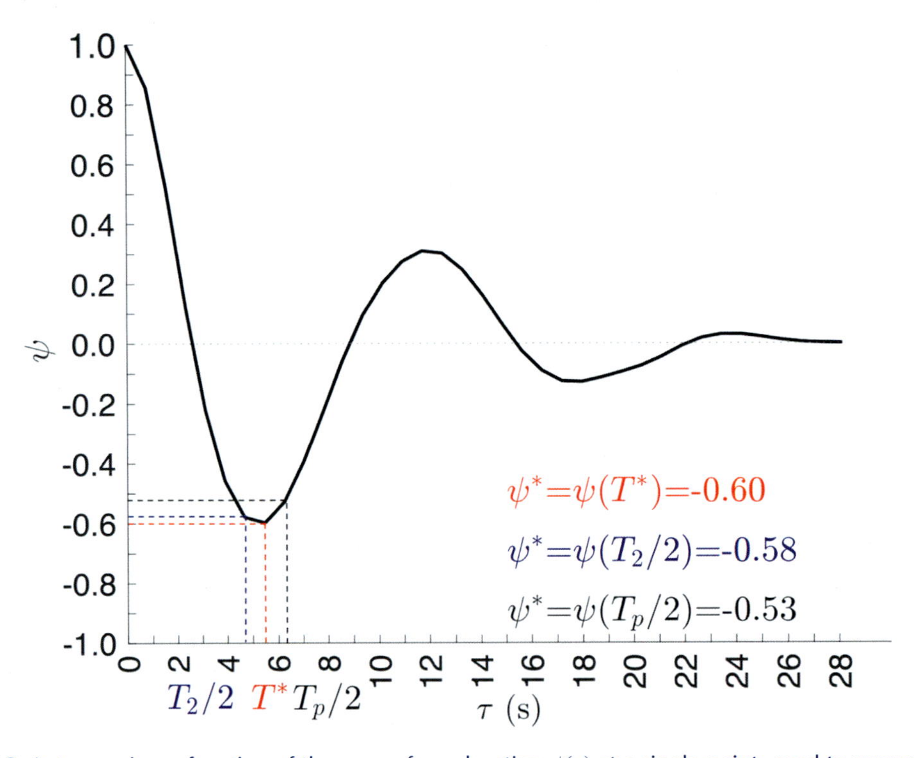

Figure 9.3 Autocovariance function of the sea surface elevation $\psi(\tau)$ at a single point, used to represent the average shape of the largest waves. As shown in the figure, ψ^* can be evaluated in different ways, for instance: (1) as the first minimum of $\psi(\tau)$, (2) at half the mean period T_2, and (3) at half the dominant period T_p. When evaluated as the first minimum of $\psi(\tau)$, the parameter ψ^* relates the crest and the trough of the largest wave in a group and is used to define $H_{\max,T}$ in the context of quasi-determinism (Boccotti, 2000) and compute $\overline{H}_{max,T}$ according to Eq. (9.7).

at different time lags τ (c_g is the group velocity). These alternatives, as shown in Fig. 9.3, can lead to differences in the largest wave height predicted, and it is therefore recommended to compute ψ^* as the minimum of the autocovariance function, as, according to this definition, it relates the crest and the trough of the largest wave in a group, in the context of the quasi-determinism theory (Boccotti, 2000). The expected maximum wave height at a single point is then obtained as:

$$\overline{H}_{max,T} = 2\sigma\sqrt{1-\psi^*}(\ln(N))^{1/2}\left(1+\frac{\gamma}{2\ln(N)}\right) \tag{9.7}$$

In their review paper on oceanic rogue waves, Dysthe et al. (2008) stated that Forristall's model for crest height and Naess's model for wave height are considered the standard model for single-point (i.e., time) extremes.

The model for the maximum envelope heights $\overline{E}_{\max,T}$ at a point is instead based on the approach originally proposed by Janssen (2003), translated for an expression for evaluating the maximum envelope height using concepts developed by Mori and Janssen (2006). The operational implementation of the freak wave warning system was presented by Janssen and Bidlot (2009). The system was further developed following Janssen (2014), leading to the second version of the system as described in Janssen (2015). Finally, Janssen (2018) recognized that the probability distribution function for large envelope wave height should have an exponential tail, which led to the introduction of the third version of the system (ECMWF 2021, Chapter 8).

Although the estimate is actually in the time domain, the width of the directional distribution, hence the short-crestedness of the sea state, is taken into account. Deviations from the normality of the probability distribution (otherwise Gaussian) occur when the spectrum is narrow both in frequency and direction. Such deviations are measured in terms of the excess kurtosis k and skewness λ of the sea surface elevation. In this context, the expected maximum envelope height at a fixed point is obtained as:

$$\overline{E}_{max,T} = \sqrt{\frac{1}{2}\langle E_{\max}\rangle}H_s \tag{9.8}$$

with

$$\langle E_{max}\rangle = \frac{1}{\beta_J}\left[G_2 - 2G_1(\alpha_J + \ln(N_J)) + \ln(N_J(2\alpha_J + \ln(N_J)))\right] \tag{9.9}$$

with $G_1 = -\gamma$, $G_2 = \gamma^2 + \pi^2/6$, where α_J and β_J were found by matching the theoretical exceedance probability for wave energy of Janssen (2014) with the expression with an exponential tail for energy level corresponding to an envelope wave height of 2.2 times the significant wave height (Janssen, 2018) to yield:

$$\beta_J = 2(\alpha_J + 1)$$

$$\alpha_J = \frac{f_b^2 - 2E_b}{2(E_b + f_b)}, f_b = \ln(P_{th}(E_b))$$

with

$$P_{th}(E_b) = e^{-E_b}\left[1 + \kappa A(E_b) + \lambda^2 B(E_b)\right]$$

$$E_b = 10$$

$$A(E_b) = \frac{1}{2}E_b(E_b - 2)$$

$$B(E_b) = \frac{1}{2}E_b(E_b^2 - 6E_b + 6)$$

The excess kurtosis κ for both free and bound waves and the skewness λ of the sea surface elevation are obtained from the wave spectrum using approximations derived in Janssen (2018) and implemented in ECWAM as given in ECMWF (2021) (Chapter 8).

Finally, the number of events N_J depends on the expectation value of E_{max} and is therefore obtained by iterations:

$$N_J = N_{slc}\sqrt{\langle E_{\max}\rangle/2}$$

where N_{slc} is an estimate of the number of wave groups in a time series of duration Γ:

$$N_{slc} = 2\nu\omega_m\Gamma/\sqrt{2\pi} \tag{9.10}$$

where $\omega_m = m_{001}m_{000}^{-1}$ is a mean angular frequency, while $\nu = (m_{000}m_{002}/m_1^2 - 1)^{1/2}$ is the spectral bandwidth parameter.

Space-time extremes

It has been shown that the maximum crest and crest-to-trough heights over a sea surface area are generally larger than those attained at a single point within the area (Barbariol et al., 2015; Benetazzo et al., 2015; Dysthe et al., 2008; Fedele et al., 2013; Krogstad et al., 2004). This is particularly true in short-crested conditions, typical of sea storms. Hence, to account for the short-crestedness and the 3D geometry of very large waves (Benetazzo et al., 2017; Fedele et al., 2013), a space-time extreme model extending the time-variate estimate of wave maxima to space-time is considered.

In space-time wave statistics, the sea surface elevation is regarded as a 2D space (x,y) and time t random field $\eta(x,y,t)$, and the probability distribution of $C_{max,ST}$ is related to the probability of the global maxima $\eta_{\max,ST}$ of the random field (Adler and Taylor, 2007; Piterbarg, 1996). Indeed, if the area considered is large enough compared to the characteristics of the sea state (namely, mean wavelength and mean wave crest length), it follows that $C_{max,ST} \sim \eta_{max,ST}$. The contribution of a weakly nonlinearity of the field is included in the statistics by means of the wave steepness parameter μ that describes the weight of second-order bound modes, which are dominant in stormy sea states (Fedele et al., 2016). A measure of μ in deep waters that accounts for bandwidth effects has been proposed by Fedele and Tayfun (2009) and is obtained from spectral integration as:

$$\mu = \sigma\omega_m^2(1 - \nu + \nu^2)g^{-1} \tag{9.11}$$

where ω_m and ν are, respectively, mean angular frequency and the spectral bandwidth parameter as in Eq. (9.10). In this context, the probability $P(C_{max,ST} > c)$ that the largest nonlinear crest height over an area A and a duration Γ is larger than a threshold c can be written as:

$$P(C_{max,ST} > c) = \left(2\pi N_3 \frac{c_0^2}{\sigma^2} + \sqrt{2\pi} N_2 \frac{c_0}{\sigma} + N_1\right) exp\left(\frac{c_0^2}{2\sigma^2}\right) \tag{9.12}$$

where c_0 is the linear threshold, $c = c_0 + 0.5\mu\sigma^{-1}c_0^2$ according to Tayfun (1980), and N_3, N_2, and N_1 are the average number of waves in the space-time volume $A\Gamma$, on its faces, and on its borders, respectively. The number of waves depends, on the one hand, upon the sea surface area $A = XY$ and the duration Γ and, on the other hand, upon the characteristics of the sea state (namely, mean wavelength L_x, mean wave crest length L_y, mean period T_2, and wave kinematics parameters). Details on how to explicitly compute the above-mentioned parameters from the wave spectrum can be found in Fedele (2012) and Benetazzo et al. (2017). Note that if $X = Y = 0$, the time extreme estimate of the expected maximum crest height based on Tayfun (1980) is obtained. This approach, together with the time extreme Janssen (2018) method, was employed to interpret the observations of one of the largest and most iconic freak waves ever measured, the "Draupner wave" (also called "New Year wave"; Cavaleri et al., 2017; Haver, 2004).

The expected maximum crest height $\overline{C}_{max,ST}$ over a sea surface area and a given duration is approximated as:

$$\overline{C}_{max,ST} = \xi_0 + \frac{\mu}{2}\frac{\xi_0^2}{\sigma} + \sigma\gamma\left[\left(1 + \mu\xi_0\sigma^{-1}\right)\left(\xi_0\sigma^{-1} - \frac{4\pi N_3\xi_0\sigma^{-1} + \sqrt{2\pi}N_2}{2\pi N_3\xi_0^2\sigma^{-2} + \sqrt{2\pi}N_2\xi_0\sigma^{-1} + N_1}\right)^{-1}\right] \tag{9.13}$$

where ξ_0 is the mode of the EDF of the linear crest (Fedele, 2012).

The estimate of the maximum wave height in space-time is obtained by means of the quasi-determinism theory (Boccotti, 2000), which provides the average shape of the largest linear waves and, thus, couples the crest and crest-to-trough heights through the narrow-bandedness parameter ψ^*. Hence, while a probability of exceedance of maximum wave heights in space-time is not defined, the expected maximum wave height $\overline{H}_{max,ST}$ over a sea surface area and a given duration is obtained by scaling the linear crest height in space-time as (Benetazzo et al., 2017):

$$\overline{H}_{max,ST} = \left[\xi_0 + \sigma\gamma\left(\xi_0\sigma^{-1} - \frac{4\pi N_3\xi_0\sigma^{-1} + \sqrt{2\pi}N_2}{2\pi N_3\xi_0^2\sigma^{-2} + \sqrt{2\pi}N_2\xi_0\sigma^{-1} + N_1}\right)^{-1}\right]\sqrt{2(1-\psi^*)} \tag{9.14}$$

If $X = Y = 0$, the time extreme estimate of the expected maximum wave height based on Boccotti (2000) is obtained.

Assessment against observations

Pivotal for the extreme wave prediction from spectral wave models is the assessment of model outputs against observations. This operation typically requires sea surface elevation η datasets. These are gathered (1) from single-point instruments as wave buoys, probes, or single-beam radars for comparison with time extreme predictions (hence, $\eta = \eta(t)$), and (2) from instruments framing portions of the sea surface as stereo-video imagery, X-band radars, and lidars for comparison with space-time extreme predictions ($\eta = \eta(x,y,t)$). While the former routinely operate at several coastal and open-ocean stations, also covering the long term, the latter are usually deployed upon requests to collect data during short-term *ad hoc* experiments. It follows that time extreme assessment has been much more extensive than space-time extreme.

Sea surface elevation datasets are processed in order to provide empirical estimates of the expected value of maximum crest, crest-to-trough, and envelope heights. Usually, the average among different realizations of the same sea state is used as a proxy of the expected value. The different realizations are obtained by splitting the dataset into sub-sets sharing the same wave spectrum, provided the sea state remains stationary and homogeneous. Given the randomness of the process, it is quite crucial to compare expected values of model maxima with consistent observations, avoiding the use of variables like the individual H_{max}, often measured by wave buoys, which represent a single realization of the random process.

In this context, some of the best practices that can be employed to properly assess the modeled wave maxima are worth mentioning. Barbariol et al. (2019) arranged a data processing methodology to obtain empirical estimates of the time and space-time extremes, respectively, from buoy and stereo-video imagery observations. For single-point sea surface elevation data $\eta(t)$, after a quality check (excluding spikes and outliers affecting the estimate of maxima), the following procedure, in agreement with Janssen and Bidlot (2009), can be applied:

- The whole time series $\eta(t)$ is partitioned into segments, i.e., sea states, of short duration (1−3 hours), that are assumed to be stationary and represented by the same wave spectrum.
- Each segment is further partitioned into n subsegments, which represent Γ-long realizations $\eta_i(t)$ of the sea state ($i = 1,\ldots n$).
- The zero-crossing analysis is performed for each subsegment in order to extract the maximum crest C_{max_i} and crest-to-trough height H_{max_i}.
- The envelope ρ_i of each realization $\eta_i(t)$ is computed by taking the complex magnitude of the Hilbert transform of $\eta_i(t)$ and doubled to obtain the envelope height $E = 2\rho_i$. Then, the maximum of each envelope height E_{max_i} is extracted.
- The sea-state estimate of maximum crest $\overline{C}_{max}$, crest-to-trough $\overline{H}_{max}$, and envelope height $\overline{E}_{max}$ expected in a Γ-long duration is obtained by averaging over the n realizations.
- Maxima are scaled (typically upscaled) to take into account the sampling bias affecting low sampling frequency measurements (especially from wave buoys; Tayfun, 1993).

Quality-checked space-time datasets $\eta(x,y,t)$ are processed in a different way. First, $\eta(x_i,y_i,t)$ time series at every point (x_i,y_i) in space can be low-pass filtered in order to remove the high-frequency noise and taken to null mean by subtracting the time series average. Then, the following procedure is applied:

- The whole observed space-time sea surface elevation volume is partitioned into space-time sub-volumes $XY\Gamma$ that represent realizations of the same sea state; the space area is divided into n equivalent XY-wide areas, and the duration is divided into m Γ-long segments, thus generating nm independent realizations.
- The maximum sea surface elevation from each realization is extracted and labeled as C_{max_i}, after verifying that they all belong to different waves (time between two C_{max_i} should exceed at least a dominant wave period T_p; if one crest is closer to another, it should be discarded).
- Zero-crossing analysis is performed on all the time series of the sub-volume, and the maximum crest-to-trough height is labeled as H_{max_i}.
- The estimate of expected maximum crest and crest-to-trough heights in the Γ-long and XY-wide sea states are obtained by averaging the nm realizations of maxima.
- A sampling bias correction can be applied if the sampling intervals in space and time are not small enough (compared to mean wave periods and wavelengths) to strongly reduce the wave height underestimation due to equidistant sampling (Tayfun, 1993; Davison et al., 2022).

The assessment of extreme wave model predictions has involved the validation of the theoretical models implemented in spectral wave models: expected maxima from statistical models fed with wave parameters from measured spectra have been compared against empirical expected maxima (Tayfun and Fedele, 2007). Barbariol et al. (2019) showed that the Forristall, Naess, and Janssen (2018 version) models presented in Section 2.1 provide, respectively, predictions of $\overline{C}_{max,T}$, $\overline{H}_{max,T}$, and $\overline{E}_{max,T}$ in excellent agreement (mean error between -0.07 and 0.11 m) with the empirical maxima observed from a Pacific Ocean wave buoy over 4.5 years (Station Papa). Regarding space-time extremes, using stormy wave fields observed by stereo-video imagery, the agreement of theoretical and empirical expectations of crest and wave maxima (including the shape of the largest waves) has been shown after several experiments in enclosed basins as the Adriatic and Black Seas, as well as in the open Yellow Sea and Pacific Ocean (Fedele et al., 2013; Benetazzo et al., 2015, 2016, 2017, 2021b; Barbariol et al., 2019).

Extreme wave predictions obtained using model spectra or as direct model outputs have been validated in many studies, starting from Janssen and Bidlot (2009) with the validation of the ECMWF "freak wave warning system" products against Canadian and Norwegian wave buoy data. The operational ECMWF wave spectra have also been used to successfully compare the expected largest crest height from space-time extreme wave models in Eq. (9.13) with stereo-video imagery observations from an offshore platform in the Yellow Sea during tropical storm conditions (Benetazzo et al., 2021b; Fig. 9.4). In Fig. 9.4, the relative model–observation error on the normalized $\overline{C}_{max,ST}$ and $\overline{H}_{max,ST}$ (to take into account the spectral model error, at least on H_s) is smaller than 7% (absolute value). Further, the ECMWF reanalysis wave spectra, employed by Barbariol et al. (2019) to produce a model-based climatology of wave maxima at a global long-term scale, yielded time extreme wave predictions in fair agreement with empirical estimates observed at a Pacific Ocean station (Station Papa; Fig. 9.5): error metrics in the figure, as the mean error (between -0.32 and 0.87 m), take into account both the theoretical and the numerical wave model errors.

Extreme output variables of operational WW3 and WAM models have been compared to long-term single-point empirical time series of $\overline{C}_{max,T}$ and $\overline{H}_{max,T}$ in the Adriatic Sea (Acqua Alta tower) and in the North Sea (Westerland and Helgoland stations; Benetazzo et al., 2021a), spanning durations from months to years, and showing a fair agreement with observations of the spectral wave model implementations using Forristall, Naess, and Tayfun models (mean error between -0.20 and -0.08 m). To validate space-time extreme model outputs of WW3 and WAM, stereo-video imagery observations from the Acqua Alta in the northern Adriatic Sea of stormy sea states and largest crest and wave heights generating therein have been successfully employed (Barbariol et al., 2017; Benetazzo et al., 2021a).

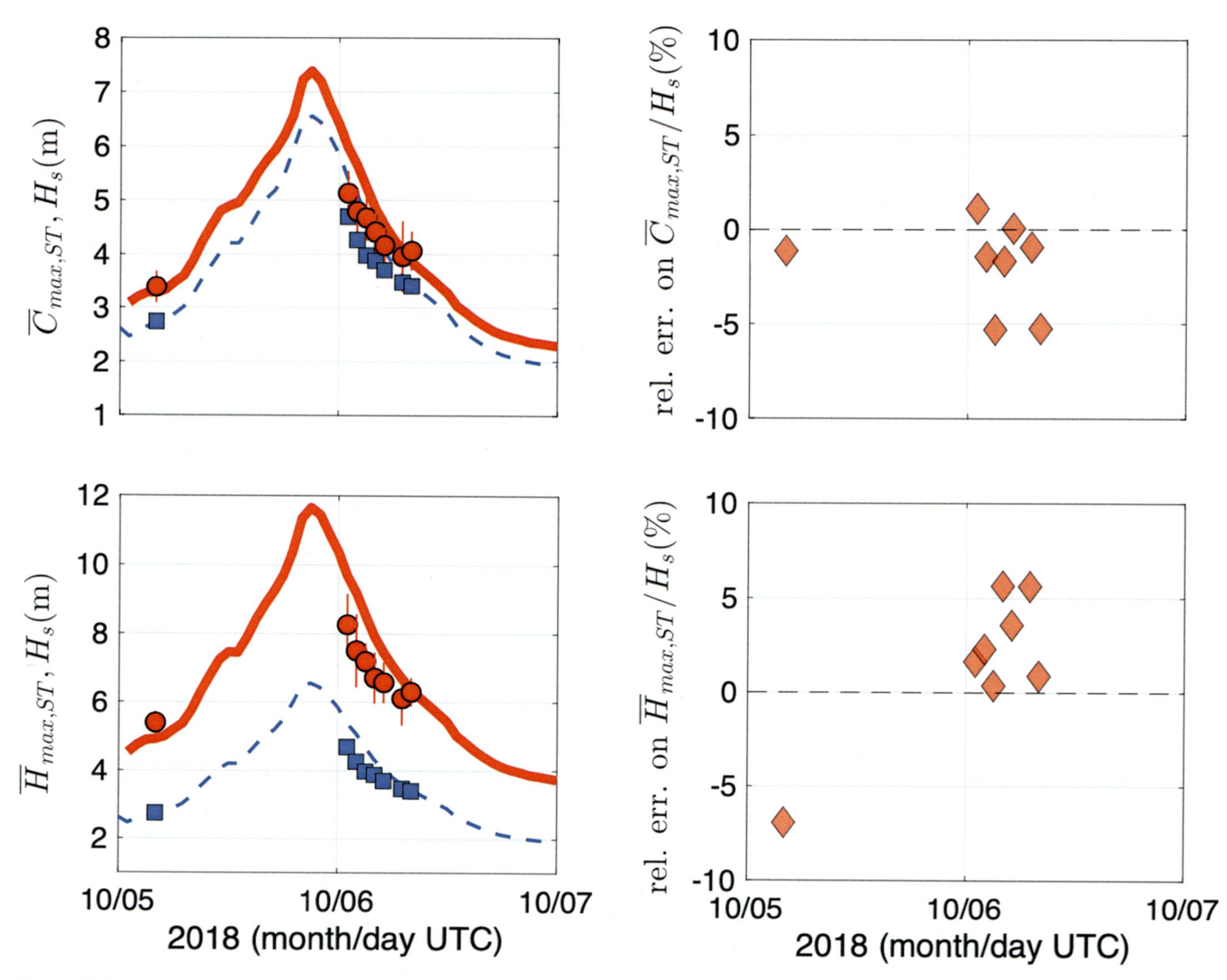

Figure 9.4 Assessment of space-time extreme prediction using ECMWF operational forecast spectra against empirical estimates in the Yellow Sea (stereo-video imagery at Gageocho station during tropical storm Kong-rey, 2018). Left panels: expected maximum crest (top) and crest-to-trough (bottom) heights as predicted by the model (red solid line) and observed (red circles). Modeled (blue dashed line) and observed (blue squares) significant wave heights are shown for reference. The duration Γ is 180 s, and the area A is 60×60 m^2. Right panels: relative errors of predictions with respect to observations for normalized crest (top) and crest-to-trough (bottom) heights.

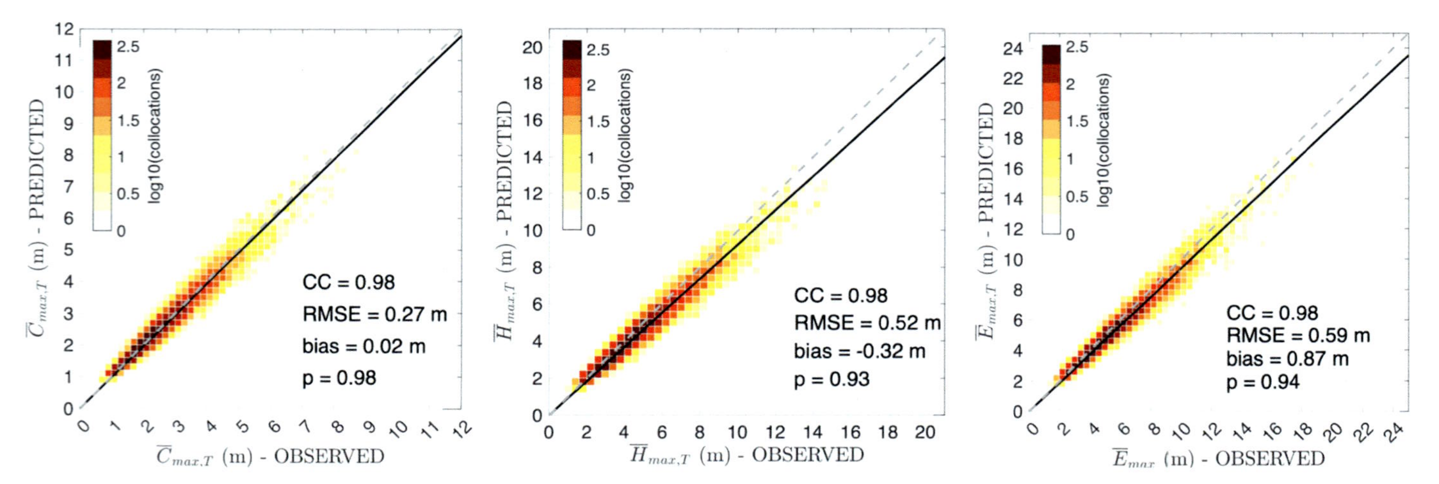

Figure 9.5 Assessment of time extremes prediction using ECMWF-ERA5 reanalysis model spectra against empirical estimates in the Pacific Ocean (Station Papa buoy, 2010–14). Left panel: expected maximum crest height from the Forristall model in 20 minutes. Central panel: expected maximum crest-to-trough height from Naess model in 20 minutes. Right panel: expected maximum envelope height from Janssen (2018) model in 20 minutes. *CC*, Correlation coefficient; *RMSE*, root mean square error; *Bias*, mean error; *p*, slope of the best fit line.

Spectral wave models

As mentioned in the previous sections, extreme wave prediction using models and equations in Section 2.1 can either (i) rely on model spectra that are outputs of wave model runs with an offline computation of the expected maxima, or (ii) be done directly from output variables of spectral wave models that are computed by *ad hoc* routines embedded in the numerical model source codes.

The wave models presently endowed with expected individual wave maxima as output variables are ECWAM, WAM, and WW3. Table 9.1 summarizes the output variables and formulations available in these models.

The expected maximum waves predicted by the statistical models presented in this chapter and embedded in spectral wave models require the domain of analysis, hence the sea state, to be stationary in time and homogeneous in space. Besides this, users should be aware that the EDFs are unbounded, as they have no physical upper bound (e.g., a limit to wave growth due to wave breaking), and they can produce unrealistic maximum waves if the domain (so, the number of waves) is too large. Therefore, in choosing as input to the model the size of the area XY and the duration Γ, which are often application-dependent, the recommendation should be to verify that the above requirements are fulfilled and that predicted heights do not violate the breaking limits. Practically, this can be attained by considering $\Gamma < [1,3]$ hours (the typical 20-minute buoy record being a good value to use), depending on the rapidity of sea-state evolution, the presence of currents, and other factors. In space, Benetazzo et al. (2020) showed that the unbounded probability distributions are reliable over surface areas with side sizes smaller than $O(10^2\ \text{m})$ for all sea states and time duration shorter than 1 hour.

Table 9.1 Time and space-time extreme output variables available in spectral wave models ECWAM, WW3, and WAM.

	Time extremes			Space-time extremes	
Model	$\overline{C}_{max,T}$	$\overline{H}_{max,T}$	$\overline{E}_{max,T}$	$\overline{C}_{max,ST}$	$\overline{H}_{max,ST}$
ECWAM	Eq. (9.4)—default Eq. (9.13) with $X = Y = 0$	Eq. (9.7)—default Eq. (9.14) with $X = Y = 0$	Eq. (9.8)	Eq. (9.13)	Eq. (9.14)
WW3	Eq. (9.13) with $X = Y = 0$	Eq. (9.14) with $X = Y = 0$	n/a	Eq. (9.13)	Eq. (9.14)
WAM	Eq. (9.4)—default Eq. (9.13) with $X = Y = 0$	Eq. (9.7)—default Eq. (9.14) with $X = Y = 0$	n/a	Eq. (9.13)	Eq. (9.14)

n/a, Not available.

Applications

Different sources of wave spectra can be used to predict wave extremes in different applications. Here, we will show examples ranging from the short-term prediction at the forecasting scale to the long-term prediction at the climatological scale.

Short-term prediction

Global forecast charts produced by the operational ECWAM-IFS system demonstrate the extreme wave prediction capabilities of spectral wave models in an operational marine prediction context. Indeed, the expected maxima are direct outputs of the ECWAM model, in agreement with Table 9.1. Even though the operational ECWAM-IFS system only offers the expected maximum envelope height alongside many other wave parameters, it can be run to release other time and space-time extreme parameters (Table 9.1). Fig. 9.6 shows the significant height and both maximum crest estimates for a 38-hour high-resolution forecast (9 km) initialized from ECMWF operational analysis, valid on November 30, 2022, at 2 UTC. Similarly, Fig. 9.7 presents the corresponding estimates for the maximum envelope and crest-to-trough heights. Besides showing the synoptic patterns of wave extremes, that substantially mirror those of the significant wave height, these charts provide a quantification of the height of the largest waves, which might be highly relevant for ship routing and for issuing warnings for the sake of maritime and offshore safety.

Long-term prediction

Hindcast and reanalysis datasets are used to simulate the wave climate over the past decades. Using the wave spectra from such datasets allows the reproduction of the climatology of wave extremes, such as the expected maximum crest, crest-to-trough, and envelope heights. First, the wave maxima are estimated at each sea state in the past; then, a statistical analysis of the whole dataset or its subsamples (e.g., on year-long chunks) is carried out at each geographical location to provide statistics of the wave maxima, such as the average, the typical value (i.e., the 50th percentile), and the extreme values (e.g., the 99th percentile). To this end, *ad hoc* open-access software have been developed in the context of projects and collaborations among scholars, such as the Coordinated Ocean Wave Climate Project (COWCLIP; https://cowclip.org/data-access/; last visit January 23, 2023).

Wave spectra can be obtained by wave model simulations purposely set up and run on domains and periods of interest, or downloaded from data repositories collecting the results of hindcasts and reanalyses (usually at a global scale),

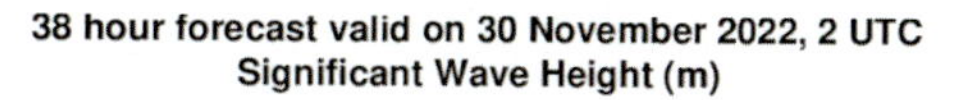

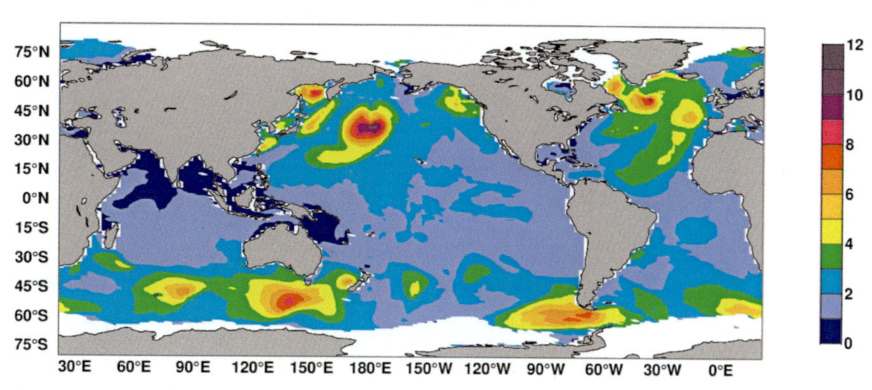

38 hour forecast valid on 30 November 2022, 2 UTC
Time Maximum Expected Crest Height (m)

75°N
60°N
45°N
30°N
15°N
0°N
15°S
30°S
45°S
60°S
75°S
30°E 60°E 90°E 120°E 150°E 180°E 150°W 120°W 90°W 60°W 30°W 0°E
18
15
12
9
6
3
0

38 hour forecast valid on 30 November 2022, 2 UTC
Space Time Maximum Expected Crest Height (m)

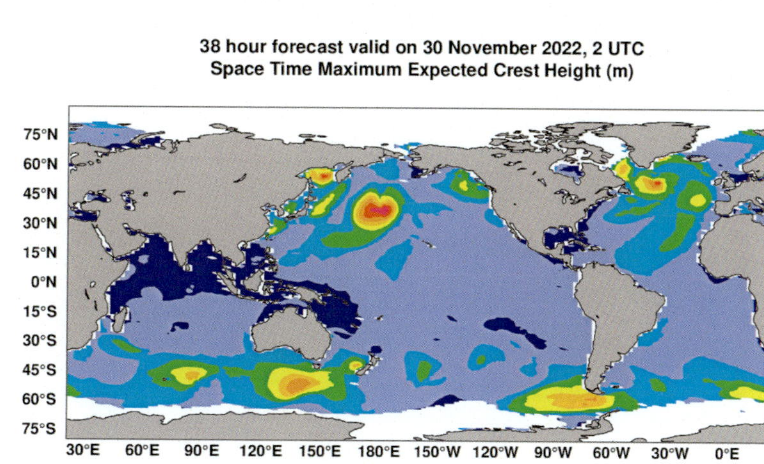

Figure 9.6 ECWAM-IFS forecast of significant wave height (top), time expected maximum crest height $C_{max,T}$ (middle) for $\Gamma = 100T_2$, and space-time expected maximum crest height $\overline{C}_{max,ST}$ (bottom) for $\Gamma = 100T_2$ and area $XY = L_x L_y$. The forecast was initialized from ECMWF high-resolution (9 km) operational analysis.

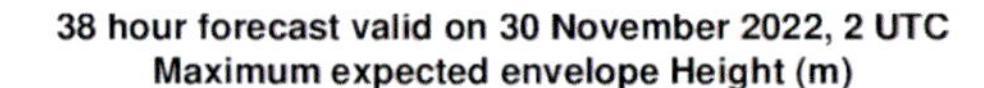

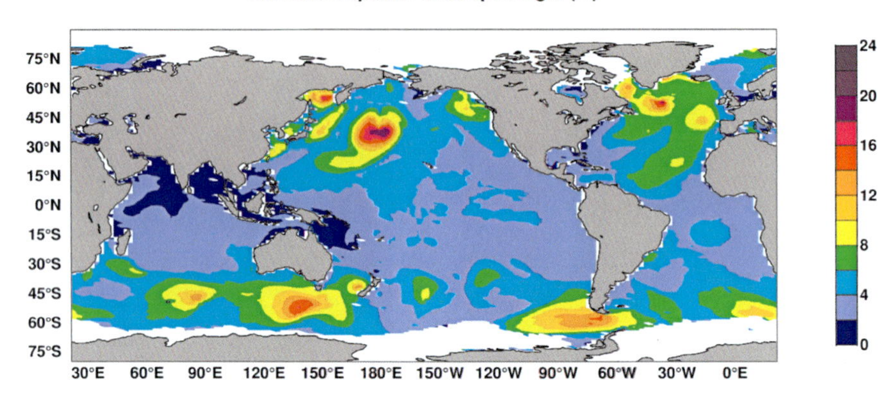

38 hour forecast valid on 30 November 2022, 2 UTC
Time Maximum Expected Crest_to_Trough Height (m)

75°N
60°N
45°N
30°N
15°N
0°N
15°S
30°S
45°S
60°S
75°S
30°E 60°E 90°E 120°E 150°E 180°E 150°W 120°W 90°W 60°W 30°W 0°E
24
20
16
12
8
4
0

Figure 9.7 ECWAM-IFS forecast of expected maximum envelope height $\overline{E}_{max,T}$ (top) for $\Gamma = 1200s$, time expected maximum crest-to-trough height $\overline{H}_{max,T}$ (middle) for $\Gamma = 100T_2$, and space-time expected maximum crest-to-through height $\overline{H}_{max,ST}$ (bottom) for $\Gamma = 100T_2$ and area $XY = L_xL_y$. The forecast was initialized from ECMWF high-resolution (9 km) operational analysis.

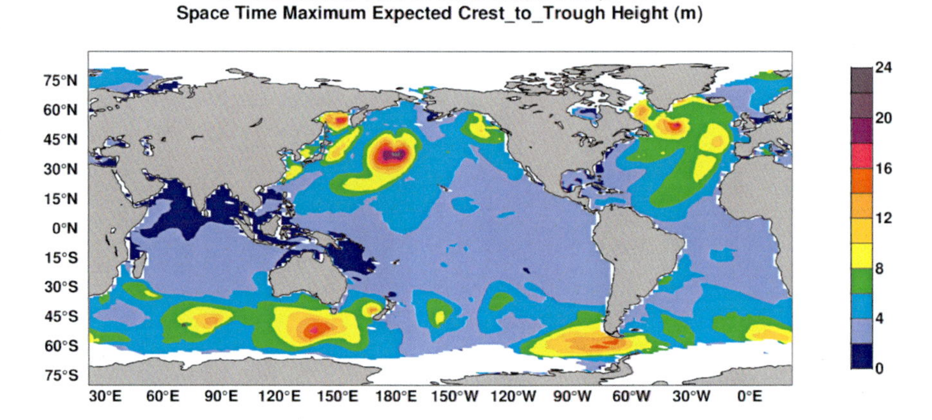

such as the Climate Data Store developed in the context of the EU Copernicus system (https://cds.climate.copernicus.eu/#!/home; last visit January 23, 2023).

As the first example, we show the average annual 99th percentile of the expected maximum crest-to-trough height obtained over the world oceans, using the spectra from an ERA5 reanalysis wind-forced global ECWAM hindcast (2000–19). Barbariol et al. (2019) achieved similar results using the previous ECMWF reanalysis ERA-Interim. $\overline{H}_{max,T}$ and $\overline{H}_{max,ST}$ have been estimated according to Eq. (9.7) and Eq. (9.14) following a time extreme and a space-time extreme approach, respectively, and then processed with the COWCLIP statistical toolbox. The maps in Figs. 9.8 and 9.9, with data at 1×1 degrees resolution, allow the identification of the regions where the largest wave heights occur and the quantification of their expected height, for the sake of offshore structure design purposes, for instance. Besides, comparing the two allows us to appreciate the difference between the time extremes and the space-time extremes, which in the example shown here can be even 50% larger.

Fig. 9.10 provides an example of $\overline{H}_{max,ST}$ long-term prediction on a regional scale. The average winter 99th percentile in the Mediterranean Sea has been obtained after the statistical analysis of the largest wave height directly produced by the WW3 model (version 6.07) forced by the ERA5 reanalysis winds in the 2000–19 period, as in Barbariol et al. (2021a). Again, the map (5 km × 5 km resolution) highlights the climatological features of the region, period, and season considered, supporting designers and decision-makers in their activities. Similar statistics were produced for the Black Sea with WAM, and the results can be found in von Schuckmann et al. (2021).

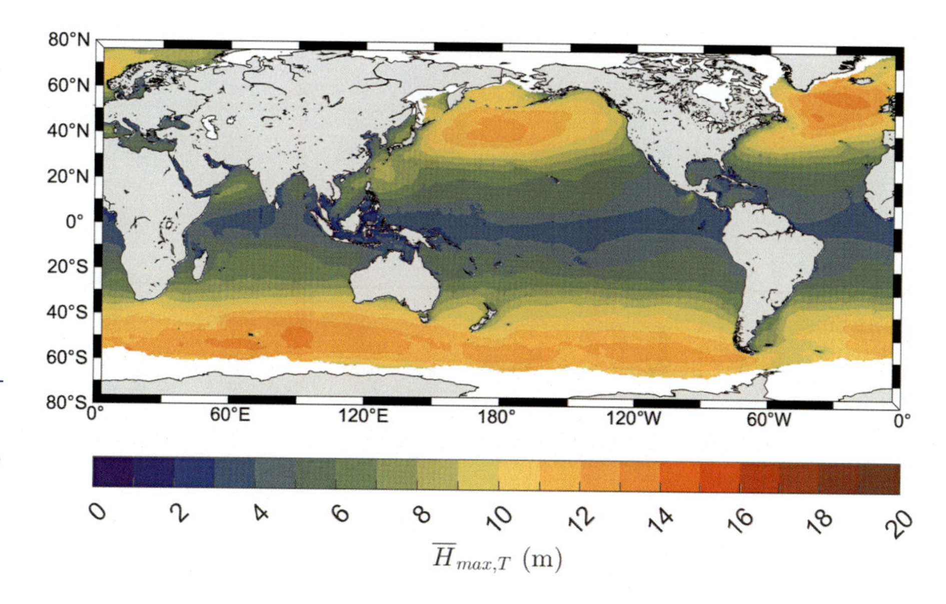

Figure 9.8 Average annual 99th percentile of the time expected maximum crest-to-trough height $\overline{H}_{max,T}$ (time extremes) over the global oceans (2000–19). Duration $\Gamma = 100T_2$.

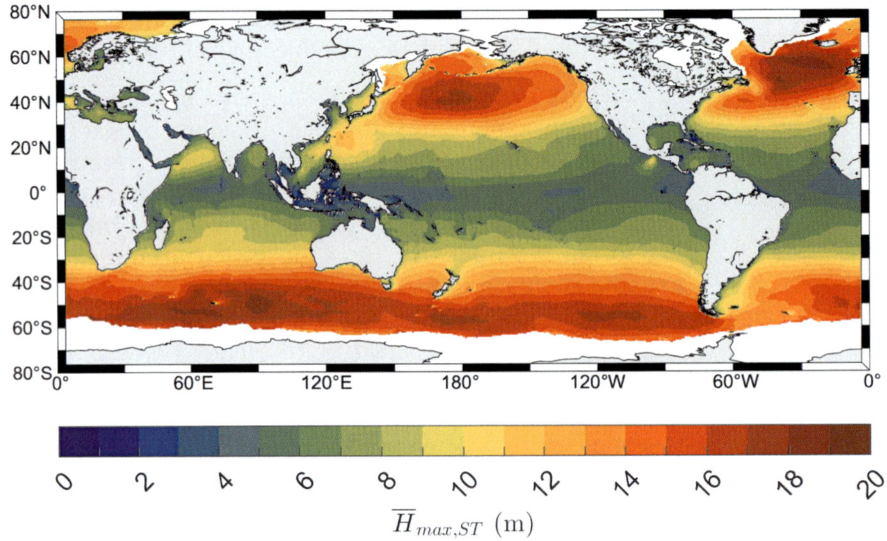

Figure 9.9 Average annual 99th percentile of the space-time maximum expected crest-to-trough height $\overline{H}_{max,ST}$ over the global oceans (2000–19). Duration $\Gamma = 100T_2$, area $XY = L_xL_y$.

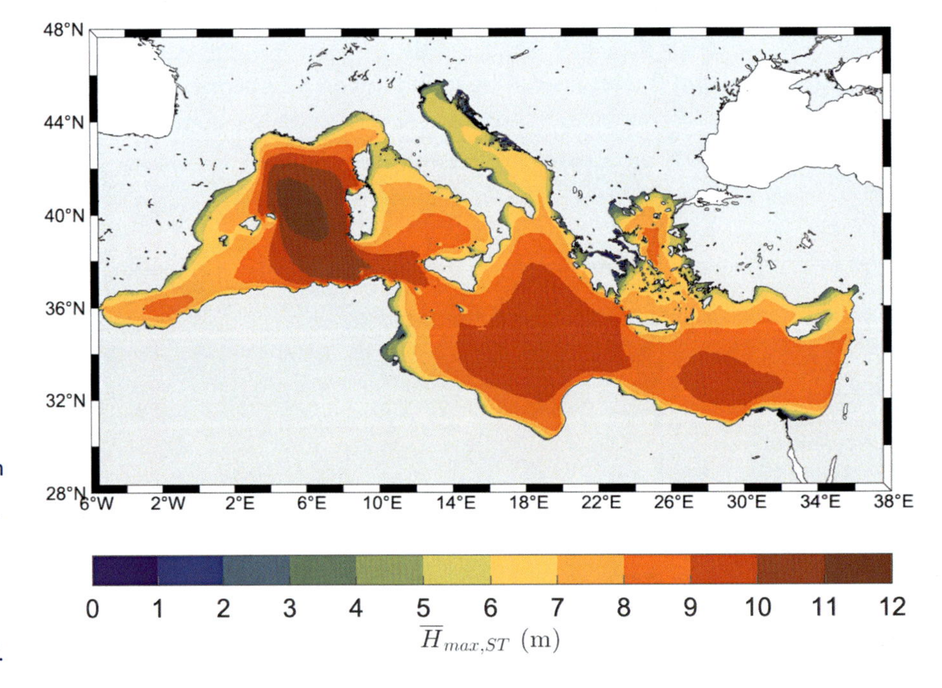

Figure 9.10 Average winter 99th percentile of the space-time expected maximum crest-to-trough height $\overline{H}_{max,ST}$ over the Mediterranean Sea (2000–19). Duration $\Gamma = 1200$ s, area $XY = 100 \times 100$ m^2.

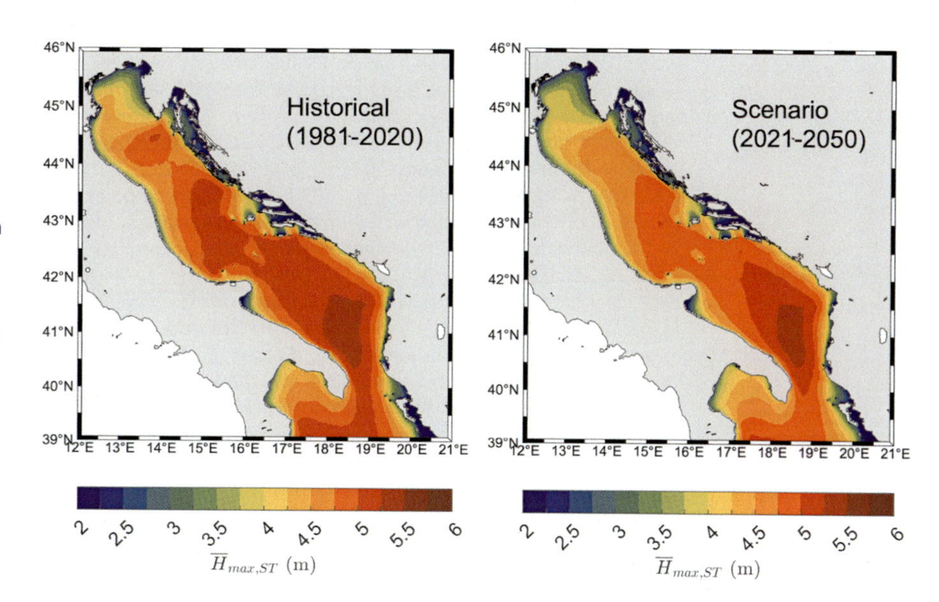

Figure 9.11 Average annual 99th percentile of the space-time expected maximum crest-to-trough height $\overline{H}_{max,ST}$ over the Adriatic Sea in historical (left panel; 1981–2010) and scenario (right panel; 2021–50, according to IPCC-RCP8.5 emission scenario) runs. Duration $\Gamma = 1200$ s, area $XY = 6 \times 6$ m^2.

Finally, we show that wave spectra from future scenario simulations can be used to project the climate of wave maxima to the near or far future. In Fig. 9.11, we compare the average annual 99th percentile of the space-time maximum expected crest-to-trough height $\overline{H}_{max,ST}$ over the Adriatic Sea in a historical run (left panel; 1981–2010) and a scenario run (right panel; 2021–50). Extreme wave heights are produced directly as outputs of the WW3 model (version 6.07), forced by ERA5 reanalysis winds, statistically corrected to locally match the wind climate produced by the COSMO-CLM high-resolution climatological model in the past and in the future, according to the IPCC-RCP8.5 emission scenario (refer to Benetazzo et al. (2022) for a description of the methodology used for the runs). By doing so, the climate change effects on the wave maxima can be studied and assessed (Barbariol et al., 2021b).

References

Adler, R.J., Taylor, J.E., 2007. Random Fields and Geometry. Springer, p. 115.

Barbariol, F., Alves, J.-H.G.M., Benetazzo, A., Bergamasco, F., Bertotti, L., Carniel, S., et al., 2017. Numerical modeling of space-time wave extremes using WAVEWATCH III. Ocean Dynamics 67, 535–549. Available from: https://doi.org/10.1007/s10236-016-1025-0.

Barbariol, F., Benetazzo, A., Carniel, S., Sclavo, M., 2015. Space-time wave extremes: the role of metocean forcings. Journal of Physical Oceanography 45 (7), 1897–1916. Available from: https://doi.org/10.1175/JPO-D-14-0232.1.

Barbariol, F., Bidlot, J.-R., Cavaleri, L., Sclavo, M., Thomson, J., Benetazzo, A., 2019. Maximum wave heights from global model reanalysis. Progress in Oceanography 175 (November 2018), 139–160. Available from: https://doi.org/10.1016/j.pocean.2019.03.009.

Barbariol, F., Davison, S., Falcieri, F.M., Ferretti, R., Ricchi, A., Sclavo, M., et al., 2021a. Wind waves in the Mediterranean Sea: an ERA5 reanalysis wind-based climatology. Frontiers in Marine Science 8. Available from: https://doi.org/10.3389/fmars.2021.760614.
Barbariol, F., Davison, S., Mercogliano, P., Sclavo, M., Benetazzo, A., 2021b. Adriatic Sea wind-wave climate years 1981–2010 and 2021–2050 (RCP8.5) (1.0) [Data set]. Zenodo . Available from: https://doi.org/10.5281/zenodo.4751163.
Benetazzo, A., Barbariol, F., Davison, S., 2020. Short-term/range extreme-value probability distributions of upper bounded space-time maximum ocean waves. Journal of Marine Science and Engineering 8 (9), 679. Available from: https://doi.org/10.3390/JMSE8090679.
Benetazzo, A., Barbariol, F., Bergamasco, F., Bertotti, L., Yoo, J., Shim, J.-S., et al., 2021b. On the extreme value statistics of spatio-temporal maximum sea waves under cyclone winds. Progress in Oceanography 197, 102642. Available from: https://doi.org/10.1016/j.pocean.2021.102642.
Benetazzo, A., Barbariol, F., Bergamasco, F., Carniel, S., Sclavo, M., 2017. Space-time extreme wind waves: analysis and prediction of shape and height. Ocean Modelling 113, 201–216. Available from: https://doi.org/10.1016/j.ocemod.2017.03.010.
Benetazzo, A., Barbariol, F., Bergamasco, F., Torsello, A., Carniel, S., Sclavo, M., 2015. Observation of extreme sea waves in a space-time ensemble. Journal of Physical Oceanography . Available from: https://doi.org/10.1175/JPO-D-15-0017.1.
Benetazzo, A., Barbariol, F., Pezzutto, P., Staneva, J., Behrens, A., Davison, S., et al., 2021a. Towards a unified framework for extreme sea waves from spectral models: rationale and applications. Ocean Engineering 219.
Benetazzo, A., Davison, S., Barbariol, F., Mercogliano, P., Favaretto, C., Sclavo, M., 2022. Correction of ERA5 wind for regional climate projections of sea waves. Water 14 (10), 1590. Available from: https://doi.org/10.3390/W14101590.
Boccotti, P., 2000. Wave Mechanics for Ocean Engineering. Elsevier Science B.V, 496 pp.
Cavaleri, L., Benetazzo, A., Barbariol, F., Bidlot, J.R., Janssen, P.A.E.M., 2017. The Draupner event: the large wave and the emerging view. Bulletin of the American Meteorological Society 98, 729–735. Available from: https://doi.org/10.1175/BAMS-D-15-00300.1.
Davison, S., Benetazzo, A., Barbariol, F., Ducrozet, G., Yoo, J., Marani, M., 2022. Space-time statistics of extreme ocean waves in crossing sea states. Frontiers in Marine Science 9, 2309. Available from: https://doi.org/10.3389/FMARS.2022.1002806/BIBTEX.
DNV GL - Det Norske Veritas Germanischer Lloyd, 2019. DNVGL-RP-C205: environmental conditions and environmental loads. DNV GL Recommended Practice 1–259.
Dysthe, K., Krogstad, H.E., Müller, P., 2008. Oceanic rogue waves. Annual Review of Fluid Mechanics 40, 287–310.
ECMWF, 2021. IFS documentation Cy47r3, part VII, ECMWF. https://www.ecmwf.int/en/elibrary/81274-ifs-documentation-cy47r3-part-vii-ecmwf-wave-model.
Fedele, F., 2012. Space–time extremes in short-crested storm seas. Journal of Physical Oceanography 42 (9), 1601–1615.
Fedele, F., Tayfun, M.A., 2009. On nonlinear wave groups and crest statistics. Journal of Fluid Mechanics 620, 221–239.
Fedele, F., Benetazzo, A., Gallego, G., Shih, P.C., Yezzi, A., Barbariol, F., et al., 2013. Space-time measurements of oceanic sea states. Ocean Modelling 70, 103–115.
Fedele, F., Brennan, J., Ponce de León, S., Dudley, J., Dias, F., 2016. Real world ocean rogue waves explained without the modulational instability. Scientific Reports 6 (1), 27715. Available from: https://doi.org/10.1038/srep27715.
Forristall, G.Z., 2000. Wave crest distributions: observations and second-order theory. Journal of Physical Oceanography 38 (8), 1931–1943.
Gumbel, E.J., 1958. Statistics of Extremes. Columbia University Press, New York.
Haver, S., 2004. A possible freak wave event measured at the Draupner jacket January 1 1995. In: Rogue Waves 2004: Proceedings of a Workshop Organized by Ifremer and Held in Brest, France, pp. 1–8. http://www.ifremer.fr/web-com/stw2004/rw/fullpapers/walk_on_haver.pdf.

Janssen, P.A.E.M., 2003. Nonlinear four-wave interactions and freak waves. Journal of Physical Oceanography 33, 863–884.
Janssen, P.A.E.M., 2014. On a random time series analysis valid for arbitrary spectral shape. Journal of Fluid Mechanics 759, 236–256.
Janssen, P.A.E.M., 2015. Notes on the maximum wave height distribution. Technical Memorandum. ECMWF, Issue 755, p. 19.
Janssen, P.A.E.M., 2018. Shallow water version of freak wave warning system. Technical Memorandum. ECMWF, Issue 813, p. 40.
Janssen, P.A.E.M., Bidlot, J.-R., 2009. On the extension of the freak wave warning system and its verification. Technical Memorandum. ECMWF, Issue 588, p. 42.
Krogstad, H.E., Liu, J., Socquet-Juglard, H., Dysthe, K.B., & Trulsen, K., 2004. Spatial extreme value analysis of nonlinear simulations of random surface waves. In: Proceedings of the ASME 2004 23rd International Conference on Ocean, Offshore and Arctic Engineering.
Mori, N., Janssen, P.A.E.M., 2006. On kurtosis and occurrence probability of freak waves. Journal of Physical Oceanography 36, 1471–1483.
Naess, A., 1985. On the distribution of crest to trough wave heights. Ocean Engineering 12 (3), 221–234. Available from: https://doi.org/10.1016/0029-8018(85)90014-9.
Piterbarg, V.I., 1996. Asymptotic Methods in the Theory of Gaussian Processes and Fields. AMS Translation of Math Monographs.
Tayfun, M.A., 1980. Narrow-band nonlinear sea waves. Journal of Geophysical Research 85 (C3), 1548–1552.
Tayfun, M.A., 1993. Sampling-rate errors in statistics of wave heights and periods. Journal of Waterway, Port, Coastal, and Ocean Engineering 119 (2), 172–192. Available from: https://doi.org/10.1061/(ASCE)0733-950X(1993)119:2(172).
Tayfun, M.A., Fedele, F., 2007. Wave-height distributions and nonlinear effects. Ocean Engineering 34, 1631–1649.
The WAVEWATCH III Development Group (WW3DG), 2016. User manual and system documentation of WAVEWATCH III version 5.16. In: Technical Note, MMAB Contribution.
von Schuckmann, K., le Traon, P.Y., Smith, N., Pascual, A., Djavidnia, S., Gattuso, J.P., et al., 2021. Copernicus marine service ocean state report, issue 5. Journal of Operational Oceanography 14 (S1), 1–185. Available from: https://doi.org/10.1080/1755876X.2021.1946240.

Further reading

Janssen, P.A.E.M., Janssen, A.J., 2019. Extreme ocean waves are a transient phenomenon. Journal of Fluid Mechanics 859, 720–818.

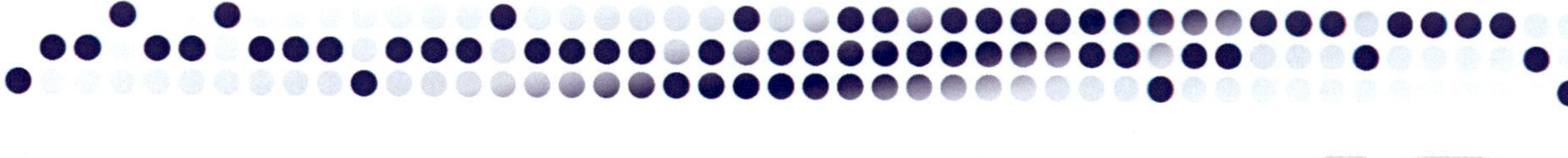

Application 1: ship responses to freak waves

Hidetaka Houtani
School of Engineering, The University of Tokyo, Bunkyo, Tokyo, Japan

Introduction

Freak waves suddenly emerge in the ocean, with heights greater than twice the significant wave height, and can cause devastating damage to ships and offshore structures. The freak wave, known as the "Draupner wave" or "New Year's wave," whose height exceeded 25 m, was encountered at an offshore gas platform in the North Sea on January 1, 1995 (Walker et al., 2004). This wave triggered many studies on the mechanism of freak wave generation and its effects on ships and offshore structures (Bitner-Gregersen and Gramstad, 2015). One of the concerns from the engineering point of view was whether the possibility of freak wave encounters should be considered during ship design. To ensure the safe operation of ships and offshore structures, it is essential to clarify the effect of freak waves on them. This chapter provides an overview of the fundamentals of ship responses to freak waves. The related experimental techniques, namely, freak wave generations in tanks and ship model design, are also studied.

Fundamentals of ship responses to freak waves

Ship responses to freak waves are complex nonlinear phenomena. Nonlinearities are present in the waves, ship motions, and loads acting on the ships. This section provides an overview of the fundamental concepts of the nonlinearities in the ship responses to freak waves. This chapter focuses mainly on longitudinal ship responses, especially those related to the hull girder strength of the ships. However, note that other potentially dangerous responses, such as lateral motions related to ship stability, can also be generated (Clauss and Hennig, 2004), which have not been considered in this chapter.

Science and Engineering of Freak Waves.
DOI: https://doi.org/10.1016/B978-0-323-91736-0.00007-9

Nonlinearity in ship motions and loads

Waves act as external forces and induce motions and loads on ships in the ocean. These ship motions and loads can be estimated by solving the equations of motion for the ships. The equation of motion for a ship in waves can be expressed as:

$$M\frac{d^2X}{dt^2} = F \tag{10.1}$$

where M, X, and F denote the mass matrix (mass and inertia moment), six-degree-of-freedom motions, and external forces acting on the ship hull from the water, respectively. F is generally decomposed into four terms with different contributions:

$$F = S + W + D + R \tag{10.2}$$

here, S represents the static restoring force corresponding to the buoyancy of the submerged volume. W is the force due to incident waves and is known as the Froude–Krylov force. D represents the diffraction force induced by the ship's body diffracting the incident waves. R denotes the radiation force, which is the reaction force from the waves, radiated owing to the ship's motion.

The loads V (e.g., vertical bending moment [VBM]) acting on the hull cross section at an arbitrary longitudinal position x is evaluated from the difference between the moment due to inertia forces, I, and the moment due to external forces, F (Eq. (10.2)) (e.g., Salvesen et al., 1970):

$$V(x) = \int I(x')dx' - \int F(x')dx' \tag{10.3}$$

Various numerical methods have been developed to evaluate the motion and cross-sectional loads of ships. Numerical methods based on the potential theory have been widely used as basic seakeeping codes. Such potential theory–based methods are classified into frequency- and time-domain solvers. Frequency-domain solvers assume that the system is linear (incident waves and ship motions are infinitesimal) and time-harmonic. Under the linear assumption, the hydrodynamic forces are evaluated over the mean wetted hull surface beneath the still water surface (Fig. 10.1A). Linear frequency-domain solvers were first developed based on the two-dimensional (2D) strip method (Salvesen et al., 1970). Two-dimensional methods evaluate the motions and loads of a ship by integrating the hydrodynamic forces acting on each cross section (strip) of the ship with respect to x (longitudinal direction). Three-dimensional (3D) panel methods were also developed to directly evaluate the hydrodynamic forces acting on a ship's 3D hull. The free-surface Green function method (Iwashita and Ohkusu, 1989) and Rankine source method (Iwashita et al., 1993) are examples of 3D methods.

However, the linear assumption with infinitesimal incident waves and ship motions is not applicable to the ship responses to freak waves. Temporal

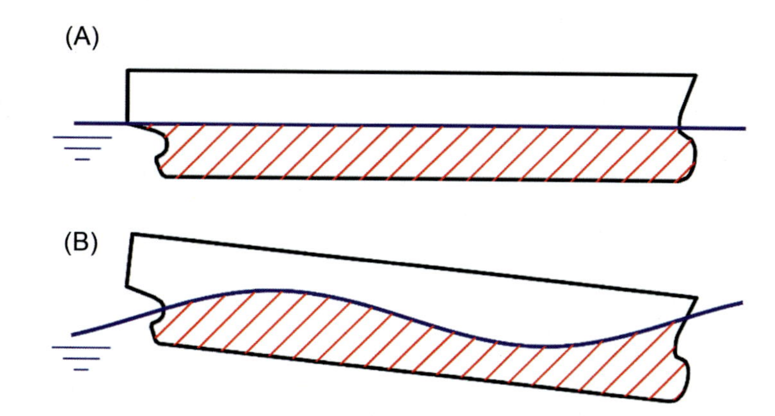

Figure 10.1 Schematic drawings of the wave surface and ship attitude considered in the (A) linear model and (B) nonlinear model. Shaded areas in red indicate the wetted hull geometry beneath the wave surface (submerged volume).

variations in the ship's attitude and wetted hull surface geometry should be considered in the analysis of large ship responses to large waves. These components are sources of nonlinearity in the ship responses and are the reasons why time-domain solvers are necessary. Various numerical methods have been developed to evaluate ship responses in large waves considering the above nonlinearities. These numerical methods are classified into the following two types according to the levels of nonlinearity considered:

1. Weakly nonlinear (Froude–Krylov nonlinear) model: Temporal variations in the ship attitudes and wetted hull surface geometry (Fig. 10.1B) are considered only for the static restoring force S and Froude–Krylov force W.
2. Weak scatter (body exact) model: In addition to (1), temporal variations in the ship attitudes and wetted hull surface geometry (Fig. 10.1B) are considered for the radiation and diffraction forces (R and D) also.

More detailed classifications can be found in Singh and Sen (2007) and ISSC (2009).

The asymmetry in the sagging and hogging moments is a well-known nonlinear response of VBMs. Such asymmetry is significant for ships with large bow flares (Watanabe et al., 1989). This nonlinear response is typically referred to as "body nonlinearity" (ISSC, 2009). This sagging/hogging asymmetry is well captured by the weakly nonlinear models (1) (Fonseca and Soares, 2002). Such weakly nonlinear seakeeping codes have been used to simulate ship responses in freak waves and have been reported to be better than the linear seakeeping codes, through tank experiments (Clauss et al., 2004a,b; Soares et al., 2006). It has also been reported that weakly nonlinear models can overestimate VBMs in large waves for ships with significant bow flares (Rajendran et al., 2016a) because of the disturbance to the waves due to the radiation and diffraction from large bow flares. Accordingly, Rajendran et al. (2016a,b) reported that a weak scatter code (2) could better predict ship responses to freak waves.

Transient vertical bending responses

The ship responses discussed in the previous subsection assume that the ship's body is rigid; therefore the timescale of these responses is the wave period. However, actual ship hulls are flexible and behave elastically, and the timescale is much shorter than the wave periods. A well-known hydroelastic response of ships to large waves is whipping, a transient structural vibration in the ship's hull girder at its natural frequency. Whipping is triggered by slamming, which is an impact phenomenon between the ship hull and wave surface. The superimposition of high-frequency whipping vibrations on the low-frequency responses at wave periods can result in significant VBMs. The occurrence of whipping vibrations and the resultant VBMs can lead to ship fracture if the VBMs exceed the hull girder's ultimate strength (Yamamoto et al., 1984). An example of the VBM of a container ship in a freak wave, measured in a tank experiment, is shown in Fig. 10.2. The high-frequency ($\sim$0.2 s) whipping vibration is triggered by slamming impacts (Fig. 10.2C) and superimposed on the low-frequency component at the wave period ($\sim$1.0 s) (Fig. 10.2B). This superimposition results in an extreme sagging moment (Fig. 10.2A).

Numerical methods to evaluate ship responses to waves, including structural vibrations, were initially developed by treating the ship as a flexible beam. The dynamic responses of structural vibrations and rigid body motions were simultaneously solved in the framework of the time-domain nonlinear strip method (Bishop and Price, 1979; Yamamoto et al., 1980). Such seakeeping codes have been widely used to analyze the hydroelastic responses of ships to large waves, including freak

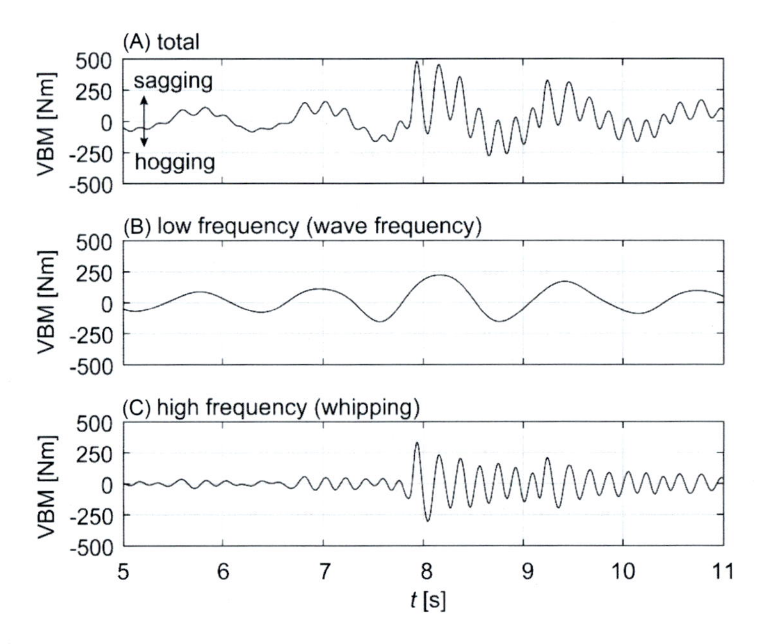

Figure 10.2 Example time series of VBM induced in a container ship by a freak wave, measured in a tank experiment. (A) Total VBM, (B) low-frequency component at the wave period, (C) high-frequency whipping component.

waves (Yamamoto et al., 1984; Kinoshita et al., 2006; Rajendran et al., 2016c). Recently, 3D panel methods have also been introduced instead of the 2D strip method (Iijima et al., 2008; Kim et al., 2015). Furthermore, with the recent advances in computational resources, various computer-aided engineering tools have been introduced to solve ship hydroelastic problems. The introduction of the finite element method addresses complicated ship structural vibration modes, and the introduction of computational fluid dynamics addresses violent fluid phenomena, including wave breaking (Takami et al., 2020a).

Nonlinearity in waves

Most studies on ship responses to freak waves have focused on the nonlinear ship responses to large waves or structural vibration phenomena, as described in the previous subsections. In most of these studies, the waves have been considered as external forces acting on ships in the framework of the linear wave theory. However, wave interactions with a third-order nonlinearity (e.g., quasi-resonant interactions) play an essential role in generating freak waves, as explained in Chapters 1, 2, and 5 of this book. In a sea state with a narrow spectral bandwidth, the probability of occurrence of freak waves is greater than that in the linear wave theory because of such nonlinear wave interactions (Chapter 5). Therefore, the statistics of the ship responses to waves can deviate from the linear estimation if nonlinear wave interactions are considered. Indeed, some recent numerical studies implied that the tails of the probability distributions of ship motions and wave-frequency VBMs were enhanced by considering the nonlinear wave interactions (Guo et al., 2019; Xiao et al., 2020; Houtani et al., 2023). In these studies, nonlinear waves precomputed by the higher-order spectral method (West et al., 1987; Dommermuth and Yue., 1987) were input to the time-domain nonlinear 2D or 3D seakeeping codes. Whether or not such wave nonlinearity affects the ship responses in a real ocean composed of directional seas is an important research topic.

Thus ship responses to freak waves are a nonlinear and transient phenomenon. The key to understanding this complex phenomenon is unraveling individually the nonlinear physics behind the waves, ship motions, and loads acting on ships.

Tank experiments on ship responses to freak waves

As discussed in **Fundamentals of ship responses to freak waves**, both the wave and ship responses are nonlinear in the case of freak waves. Tank experiments are useful for investigating such nonlinear and complicated phenomena. This section reviews the experimental techniques, such as freak wave generation in tanks and ship models, which have been developed to measure cross-sectional loads. Some experimental studies using these techniques to examine ship responses to freak waves are also detailed.

Wave generation method

Freak wave generation methods in experimental tanks based on linear wave-focusing mechanisms and the nonlinear quasi-resonant interactions are overviewed in the following subsections.

Wave generation based on linear wave-focusing mechanism

Various large wave generation methods that utilize the dispersion relation of linear waves have been proposed. In deep waters, waves propagate with variable velocities that depend on the wavelength λ or wave period T, owing to their dispersive property. From the linear dispersion relation, the phase speed c_p and group velocity c_g of the waves can be expressed as follows:

$$c_p = \sqrt{\frac{g}{k}} = \frac{g}{\omega}, \quad c_g = \frac{1}{2} c_p, \tag{10.4}$$

where $\omega\left(=\frac{2\pi}{T}\right), k\left(=\frac{2\pi}{\lambda}\right)$, and g denote the angular frequency, wavenumber, and gravitational acceleration, respectively.

The following two methods are widely used for unidirectional large wave generation (Chaplin, 1996).

Phase speed method

The phases of the component waves at the wavemaker are determined considering the linear phase speed c_p (Eq. (10.4)) such that all component waves are in phase at the target location and time (Greenhow et al., 1982). The amplitudes of each component wave are also considered if a specific large wave elevation time series is to be reproduced, such as an observed freak wave (Clauss et al., 2004a,b).

Group velocity method

A transient wave is generated such that the energy is focused at the target location and time, considering the linear wave group velocity c_g (Eq. (10.4)) (Takezawa and Hirayama, 1971; Longuet-Higgins, 1974).

However, the location and time of the wave focusing deviate from those predicted by the linear wave theory because the wave evolves nonlinearly during its propagation, especially near the focusing point. To compensate for such nonlinear effects, some modification methods such as phase correction (Chaplin, 1996), phase speed correction (Clauss and Kühhnlein, 1997), and group velocity correction (Tomita et al., 1997; Ten and Tomita, 2006) have been proposed.

These methods have been extended to generate large directional waves by considering the phase convergence for directional wave components. Using wave directionality, Ueno et al. (2013) reproduced a large wave considering the sea state when a ship accident occurred, and Schmittner and Hennig

Figure 10.3 Image of a freak wave, mimicking a woodblock print "The Great Wave off Kanagawa" by Katsushika Hokusai, generated under a crossing sea condition in a wave basin.

(2012) and McAllister et al. (2019) reproduced the Draupner wave. Fig. 10.3 shows an example image of a freak wave in a crossing sea generated in the Actual Sea Model Basin (80 m × 40 m × 4.5 m) at the National Maritime Research Institute, Japan, using this method. Two wave systems, whose mean wavelengths are 6 and 4 m, crossed at 120 degrees.

Wave generation taking into account nonlinear quasi-resonant interactions

The wave spectra do not evolve temporally or spatially within the framework of the linear wave theory. In contrast, the amplitudes of the four-wave spectral modes satisfying the quasi-resonance condition evolve temporally and/or spatially, considering the third-order nonlinearity (Zakharov, 1968). When the wave spectrum is narrow, wave modulation is enhanced because of the intensified four-wave quasi-resonant interaction, which leads to freak wave generation (Janssen, 2003; Onorato et al., 2009; Waseda et al., 2009; refer to Chapter 5). Wave generation methods that consider nonlinear quasi-resonant interactions have also been developed.

Benjamin and Feir (1967) found that the Stokes wave is unstable against modulated perturbations. This instability is known as modulational instability (MI) or Benjamin–Feir instability. MI has been investigated the most to study the evolution of water waves owing to its quasi-resonant interactions. Tulin and Waseda (1999) produced large waves in a wave tank by generating Stokes waves with modulated perturbations comprising a pair of sideband waves. This method enables the control of the focusing location and amplitude of large waves by adjusting the initial wave parameters.

The nonlinear wave evolutions of such modulated wave trains are well described by the nonlinear Schrödinger equation (NLSE) (Zakharov, 1968). The NLSE has analytical solutions, such as the Peregrine breather (PB) and Akhmediev breather, which have been reproduced in wave tanks by Chabchoub et al. (2011) and

Chabchoub et al. (2014) (and in a numerical wave tank by Ten and Tomita (2006)). A method to reproduce the arbitrary wave fields simulated by the NLSE in a tank has been proposed by Khait (2020).

The large wave generation methods reviewed so far in this subsection produce unidirectional waves. Moreover, when using the wave generation methods based on NLSE, the difference between the generated wave and NLSE prediction is large when the spectrum is broad because NLSE assumes narrow-banded spectra (Khait, 2020). To circumvent these limitations, Houtani et al. (2018a) developed a higher-order spectral method-based wave generation (HOSM-WG) technique that enabled the generation of short-crested freak waves in a wave basin without any limitations on the spectral shape (Fig. 10.4). A freak wave was detected by a Monte Carlo simulation of irregular waves with the HOSM (West et al., 1987; Dommermuth and Yue, 1987), and this freak wave was reproduced in a wave basin.

Ship model design for measuring cross-sectional loads

Ship models are used in tank experiments on ship responses to freak waves. For experiments associated with cross-sectional loads, such as the VBM, the ship models must be designed such that the cross-sectional loads can be measured. The ship models for measuring cross-sectional loads are roughly classified into the following four types (ITTC, 2014):

1. Rigid segmented model;
2. Flexible segmented model (segmented model with a backbone);
3. Hydro-structural model with a backbone; and
4. Hydro-structural model without a backbone (pure hydro-structural model).

The rigid segmented model (1) is divided into multiple segments with rigid hulls, and the cross-sectional loads are measured at each divided cross section. Adjacent segments are connected through a foundation with a load cell or flexible plates with strain gauges. The connected parts are covered with soft sheets

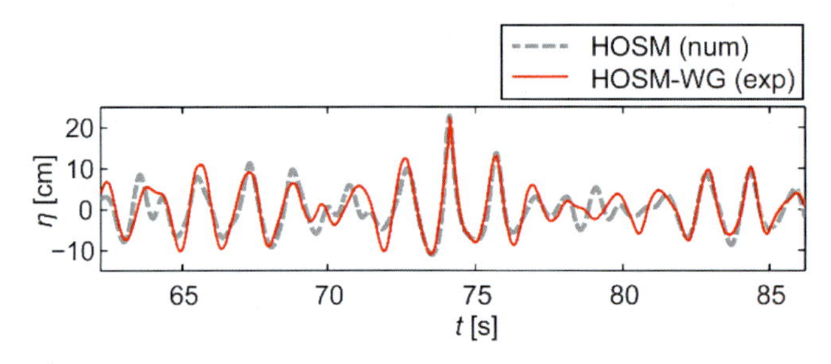

Figure 10.4 Freak wave in a directional sea generated in a wave basin using the HOSM-WG method. *Source: Redrawn from Houtani et al., 2018. Generation of a spatially periodic directional wave field in a rectangular wave basin based on higher-order spectral simulation. Ocean Engineering, 169, 429–441, Copyright (2018), with permission from Elsevier.*

to prevent the entry of water. This type of model is unsuitable for measuring structural vibrations because its stiffness is dissimilar to that of a full-scale ship.

Model (2) is also composed of multiple segments but features a flexible backbone. The material and cross-sectional shape of the flexible backbone are selected to satisfy the stiffness (mainly of the vertical bending) similarity with a full-scale subject ship. Therefore, this model is suitable for measuring cross-sectional loads, including structural vibrations. The model should be divided into more than four segments to reproduce the lowest-order (two-node) vertical bending vibration mode (Bennett et al., 2014).

Model (3) has an unsegmented continuous hull composed of a soft material, such as urethane foam, instead of segmented hulls. Both the hull and backbone are carefully designed to satisfy the stiffness similarity. Compared to the segmented model, the continuity in the hull geometry is superior from the hydrodynamic and waterproofing perspectives. However, long-term changes in the material properties of the hull are a concern. Nevertheless, the changes in the vertical bending stiffness and the influence of structural creep have been reported to be negligibly small (Houtani et al., 2019).

Model (4) does not feature a backbone, but appropriate stiffness is imparted to the model hull. This model has been developed recently to reproduce torsional vibrations similar to that of a full-scale ship (Houtani et al., 2018b) and to directly measure the stress at an arbitrary point on the hull instead of using cross-sectional loads (Komoriyama, 2022).

Towing experiments of ship models in freak waves

Many tank experiments have been performed on towing various types of ship models (**Ship model design for measuring cross-sectional loads**) while encountering freak waves generated using various methods (**Wave generation method**). For example, many studies have experimentally investigated the ship motion and load responses in observed freak waves such as the Draupner wave (Clauss et al., 2004a,b; Soares et al., 2006; Rajendran et al., 2016a). The observed freak waves were reproduced in tanks using a linear-focusing mechanism (**Wave generation based on linear wave-focusing mechanism**). Comparisons of the experimental and numerical results show the importance of weakly nonlinear and weak scatter models in evaluating the ships' responses to freak waves. These studies also compare the maximum VBMs acting on the ships during the encounters with the observed freak waves with the design loads provided by the classification society rules. In some cases, the maximum VBMs in the experiments exceeded the design values (Rajendran et al., 2016a).

Some experimental studies have also investigated ships' responses to freak waves by considering nonlinear quasi-resonant interactions. Onorato et al. (2013) and Klein et al. (2016) performed towing experiments using a rigid segmented ship model against the PB (Peregrine, 1983), which is an analytical solution to the NLSE. From the perspective of engineering applications, the ease of changing the steepness or wave height of the freak wave by adjusting the initial parameters is one of the benefits of using PB as a design wave. Houtani et al. (2019) generated modulated wave trains in a tank and towed a hydro-structural backbone ship model. VBMs composed of wave-frequency and whipping responses were investigated. It was revealed that the maximum sagging moment depended on the freak wave geometry; for a given freak wave height, larger sagging moments occurred when the rear trough was deeper than the front trough.

Discussion: are freak waves the most dangerous wave for ships?

A primary concern in ship structural design is the expected maximum dynamic load induced by waves during the lifetime of the ship. Ship structures must be designed to withstand extreme wave loads. Here, a fundamental question is raised: do freak waves cause the maximum load on ships during their lifetime?

Fonseca et al. (2006) simulated the VBM responses of a container ship to 20 freak waves observed in the ocean, including the Draupner wave, using a nonlinear strip method. Their study showed that none of the VBMs induced by these 20 freak waves exceeded the expected maximum VBM in their lifetime (at an exceedance probability of 10^{-8}) in the North Sea, estimated by the same seakeeping code. This result implies that the most dangerous wave inducing the maximum VBM on a ship during its lifetime is possibly not a freak wave.

Indeed, ship responses to waves depend on wave frequency, which is characterized by a transfer function (frequency response function [FRF]). In a linear system, the ship responses to waves are represented in the frequency domain by the product of the FRF and wave spectrum. Given such an FRF, freak waves are not necessarily expected to induce the maximum ship response. In practice, the ship response can be nonlinear and transient (**Fundamentals of ship responses to freak waves**); therefore identifying the most dangerous wave is difficult. Some methods have been proposed to estimate the most probable wave episode inducing the maximum ship response for a given wave spectrum based on the linear FRF, such as the most likely extreme response (Adegeest et al., 1998) or critical wave episode method (Kim and Kim, 2018). In these methods, nonlinear seakeeping simulations of the most probable wave episode, estimated based on the linear FRF, are performed to obtain the maximum nonlinear response. Furthermore, a method is proposed to directly estimate the most probable wave episode that induces the maximum VBM, considering the nonlinear and transient responses, by introducing the first-order reliability method (Takami et al., 2020b). Further development of such methods can help answer the question: "are freak waves the most dangerous waves for ships?" or identify the characteristics of the most dangerous waves for ships.

The effect of wave nonlinearity on ship response remains an unresolved question, although its effects on ship response statistics have been discussed in some studies (**Nonlinearity in waves**). Another possible viewpoint to be focused on for nonlinear wave characteristics, other than wave elevation statistics, might be the wave geometry. As mentioned in **Towing experiments of ship models in freak waves**, the VBM of a ship depends on the freak wave geometry (Houtani et al., 2019). The encounter timing between a ship and a freak wave significantly affects the VBM because the freak wave geometry varies significantly within a few wave periods. Moreover, Fujimoto et al. (2019) showed that the freak wave geometry was affected by its generation mechanism. Freak waves generated by linear focusing tend to be more fore-aft symmetric, whereas those generated by nonlinear quasi-resonant interactions tend to be more fore-aft asymmetric. The results of these studies imply that the ship responses may be affected by wave nonlinearity through the wave geometry.

References

Adegeest, L., Braathen, A., Vada, T., 1998. Evaluation of methods for estimation of extreme nonlinear ship responses based on numerical simulations and model tests. In: Proc. 22nd Symposium on Naval Hydrodynamics, Washington, DC, vol. 1, pp. 70–84.

Benjamin, T.B., Feir, J.E., 1967. The disintegration of wave trains on deep water Part 1. Theory. Journal of Fluid Mechanics 27 (3), 417–430.

Bennett, S.S., Hudson, D.A., Temarel, P., 2014. Global wave-induced loads in abnormal waves: comparison between experimental results and classification society rules. Journal of Fluids and Structures 49, 498–515.

Bishop, R.E.D., Price, W.G., 1979. Hydroelasticity of Ships. Cambridge University Press.

Bitner-Gregersen, E.M., & Gramstad, O., 2015. Rogue waves impact on ships and offshore structures. Det Norske Veritas Germanischer Lloyd Strategic Research and Innovation Position Paper.

Chabchoub, A., Hoffmann, N.P., Akhmediev, N., 2011. Rogue wave observation in a water wave tank. Physical Review Letters 106 (20), 204502.

Chabchoub, A., Kibler, B., Dudley, J.M., Akhmediev, N., 2014. Hydrodynamics of periodic breathers. Philosophical Transactions of the Royal Society A: Mathematical, Physical and Engineering Sciences 372 (2027), 20140005.

Chaplin, J.R., 1996. On frequency-focusing unidirectional waves. International Journal of Offshore and Polar Engineering 6 (2), 131–137.

Clauss, G.F., Hennig, J., 2004. Deterministic analysis of extreme roll motions and subsequent evaluation of capsizing risk. International Shipbuilding Progress 51 (2–3), 135–155.

Clauss, G.F., Kühhnlein, W.L., 1997. Simulation of design storm wave conditions with tailored wave groups. In: The Seventh International Offshore and Polar Engineering Conference. OnePetro.

Clauss, G.N.F., Schmittner, C.E., Hennig, J., Guedes Soares, C., Fonseca, N., Pascoal, R., 2004a. Bending moments of an FPSO in rogue waves. In: International Conference on Offshore Mechanics and Arctic Engineering, vol. 37440, pp. 455–462.

Clauss, G., Stutz, K., Schmittner, C., 2004b. Rogue wave impact on offshore structures. In: Offshore Technology Conference. OnePetro.

Dommermuth, D.G., Yue, D.K., 1987. A high-order spectral method for the study of nonlinear gravity waves. Journal of Fluid Mechanics 184, 267–288.

Fonseca, N., Soares, C.G., 2002. Comparison of numerical and experimental results of nonlinear wave-induced vertical ship motions and loads. Journal of Marine Science and Technology 6 (4), 193–204.

Fonseca, N., Soares, C.G., Pascoal, R., 2006. Structural loads induced in a containership by abnormal wave conditions. Journal of Marine Science and Technology 11 (4), 245–259.

Fujimoto, W., Waseda, T., Webb, A., 2019. Impact of the four-wave quasi-resonance on freak wave shapes in the ocean. Ocean Dynamics 69 (1), 101–121.

Greenhow, M., Vinje, T., Brevig, P., Taylor, J., 1982. A theoretical and experimental study of the capsize of Salter's duck in extreme waves. Journal of Fluid Mechanics 118, 221–239.

Guo, B., Gramstad, O., Vanem, E., Bitner-Gregersen, E., 2019. Study on the effect of climate change on ship responses based on nonlinear simulations. Journal of Offshore Mechanics and Arctic Engineering 141 (4).

Houtani, H., Komoriyama, Y., Matsui, S., Oka, M., Sawada, H., Tanaka, Y., et al., 2018b. Designing a hydro-structural ship model to experimentally measure its vertical bending and torsional vibrations. Journal of Advanced Research in Ocean Engineering 4 (4), 174–184.

Houtani, H., Matsui, S., Fujimoto, W., 2023. Numerical investigation of the statistics of vertical bending moments of ships in nonlinearly evolving irregular waves. In: Proceedings of the ASME 2023 42nd International Conference on Ocean, Offshore and Arctic Engineering.

Houtani, H., Waseda, T., Fujimoto, W., Kiyomatsu, K., Tanizawa, K., 2018a. Generation of a spatially periodic directional wave field in a rectangular wave basin based on higher-order spectral simulation. Ocean Engineering 169, 428–441.

Houtani, H., Waseda, T., Tanizawa, K., Sawada, H., 2019. Temporal variation of modulated-wave-train geometries and their influence on vertical bending moments of a container ship. Applied Ocean Research 86, 128–140.

Iijima, K., Yao, T., Moan, T., 2008. Structural response of a ship in severe seas considering global hydroelastic vibrations. Marine Structures 21 (4), 420–445.

ISSC, 2009. ISSC report of technical committee 1.2: loads. In: 17th International Ship and Offshore Structures Congress, Korea.

ITTC, 2014. Seakeeping experiments (7.5-02-07-0.02.1). ITTC—Recommended Procedures and Guidelines.

Iwashita, H., Ohkusu, M., 1989. Hydrodynamic forces on a ship moving with forward speed in waves. Journal of the Society of Naval Architects of Japan 166, 187–205 (in Japanese).

Iwashita, H., Lin, X., Takaki, M., 1993. Combined boundary-integral equations method for ship motions in waves. Transactions of the West-Japan Society of Naval Architects 85, 37–55.

Janssen, P.A., 2003. Nonlinear four-wave inter-actions and freak waves. Journal of Physical Oceanography 33 (4), 863–884.

Khait, A., 2020. Third-order generation of narrow-banded wave trains by a wavemaker. Ocean Engineering 218, 108200.

Kim, J.H., Kim, Y., 2018. Prediction of extreme loads on ultra-large containerships with structural hydroelasticity. Journal of Marine Science and Technology 23 (2), 253–266.

Kim, J.H., Kim, Y., Yuck, R.H., Lee, D.Y., 2015. Comparison of slamming and whipping loads by fully coupled hydroelastic analysis and experimental measurement. Journal of Fluids and Structures 52, 145–165.

Kinoshita, T., Shi, J., Nakasumi, S., Kameoka, H., Suzuki, K., Waseda, T., et al., 2006. Investigation of freak wave induced loads on a large container ship. 26th Symposium on Naval Hydrodynamics, 3, 211–218.

Klein, M., Clauss, G.F., Rajendran, S., Soares, C.G., Onorato, M., 2016. Peregrine breathers as design waves for wave-structure interaction. Ocean Engineering 128, 199–212.

Komoriyama, Y., 2022. A novel technique to estimate hull structural response in waves by means of hydrodynamic pressure measurement-1st report: Estimation for vertical bending moment. Journal of the Japan Society of Naval Architects and Ocean Engineers 35, (in Japanese).

Longuet-Higgins, M.S., 1974. Breaking waves in deep or shallow water, Proc. 10th Conf. on Naval Hydrodynamics, vol. 597. MIT.

McAllister, M.L., Draycott, S., Adcock, T.A.A., Taylor, P.H., Van Den Bremer, T.S., 2019. Laboratory recreation of the Draupner wave and the role of breaking in crossing seas. Journal of Fluid Mechanics 860, 767–786.

Onorato, M., Proment, D., Clauss, G., Klein, M., 2013. Rogue waves: from nonlinear Schrödinger breather solutions to sea-keeping test. PLoS One 8 (2), e54629.

Onorato, M., Waseda, T., Toffoli, A., Cavaleri, L., Gramstad, O., Janssen, P.A.E.M., et al., 2009. Statistical properties of directional ocean waves: the role of the modulational instability in the formation of extreme events. Physical Review Letters 102 (11), 114502.

Peregrine, D.H., 1983. Water waves, nonlinear Schrödinger equations and their solutions. The ANZIAM Journal 25 (1), 16–43.

Rajendran, S., Fonseca, N., Soares, C.G., 2016b. Prediction of extreme motions and vertical bending moments on a cruise ship and comparison with experimental data. Ocean Engineering 127, 368–386.

Rajendran, S., Fonseca, N., Soares, C.G., 2016c. A numerical investigation of the flexible vertical response of an ultra large containership in high seas compared with experiments. Ocean Engineering 122, 293–310.

Rajendran, S., Vásquez, G., Soares, C.G., 2016a. Effect of bow flare on the vertical ship responses in abnormal waves and extreme seas. Ocean Engineering 124, 419–436.

Salvesen, N., Tuck, E.O., Faltinsen, O., 1970. Ship motions and sea loads. Trans. SNAME 78, 250–287.

Schmittner, C., Hennig, J., 2012. Optimization of short-crested deterministic wave sequences via a phase-amplitude iteration scheme. In: International Conference on Offshore Mechanics and Arctic Engineering, vol. 44922. American Society of Mechanical Engineers, pp. 79–86.

Singh, S.P., Sen, D., 2007. A comparative linear and nonlinear ship motion study using 3-D time domain methods. Ocean Engineering 34 (13), 1863–1881.

Soares, C.G., Fonseca, N., Pascoal, R., Clauss, G.F., Schmittner, C.E., Hennig, J., 2006. Analysis of design wave loads on an FPSO accounting for abnormal waves. Journal of Offshore Mechanics and Arctic Engineering 128 (3), 241–247.

Takami, T., Iijima, K., Jensen, J.J., 2020a. Extreme value prediction of nonlinear ship loads by FORM using prolate spheroidal wave functions. Marine Structures 72, 102760.

Takami, T., Komoriyama, Y., Ando, T., Ozeki, S., Iijima, K., 2020b. Efficient FORM-based extreme value prediction of nonlinear ship loads with an application of reduced-order model for coupled CFD and FEA. Journal of Marine Science and Technology 25 (2), 327–345.

Takezawa, S., Hirayama, T., 1971. On the generation of arbitrary transient water waves. Journal of the Society of Naval Architects of Japan 1971 (129), 41–53 (in Japanese).

Ten, I., Tomita, H., 2006. Simulation of the ocean waves and appearance of freak waves. In: Reports of RIAM Symposium, No. 17SP1–2, pp. 10–11.

Tomita, H., Sawada, H., Ohmatsu, S., Yoshimoto, H., 1997. Study on dynamical and stochastic properties of breaking ocean waves. Papers of Ship Research Institute 34 (6), 313–334.

Tulin, M.P., Waseda, T., 1999. Laboratory observations of wave group evolution, including breaking effects. Journal of Fluid Mechanics 378, 197–232.

Ueno, M., Miyazaki, H., Taguchi, H., Kitagawa, Y., Tsukada, Y., 2013. Model experiment reproducing an incident of fast ferry. Journal of Marine Science and Technology 18 (2), 192–202.

Walker, D.A., Taylor, P.H., Taylor, R.E., 2004. The shape of large surface waves on the open sea and the Draupner New Year wave. Applied Ocean Research 26 (3-4), 73–83.

Waseda, T., Kinoshita, T., Tamura, H., 2009. Evolution of a random directional wave and freak wave occurrence. Journal of Physical Oceanography 39 (3), 621–639.

Watanabe, I., Ueno, M., Sawada, H., 1989. Effects of bow flare shape to the wave loads of a container ship. Journal of the Society of Naval Architects of Japan 1989 (166), 259–266.

West, B.J., Brueckner, K.A., Janda, R.S., Milder, D.M., Milton, R.L., 1987. A new numerical method for surface hydrodynamics. Journal of Geophysical Research: Oceans 92 (C11), 11803–11824.

Xiao, Q., Zhou, W., Zhu, R., 2020. Effects of wave-field nonlinearity on motions of ship advancing in irregular waves using HOS method. Ocean Engineering 199, 106947.

Yamamoto, Y., Fujino, M., Fukasawa, T., 1980. Motion and longitudinal strength of a ship in head sea and the effects of non-linearities. Journal of the Society of Naval Architects of Japan 143, 179–187.

Yamamoto, Y., Fujino, M., Ohtsubo, H., Fukasawa, T., Iwai, Y., Aoki, G., et al., 1984. 12. Disastrous damage of a bulk carrier due to slamming. Naval Architecture and Ocean Engineering 22, 159–169.

Zakharov, V.E., 1968. Stability of periodic waves of finite amplitude on the surface of a deep fluid. Journal of Applied Mechanics and Technical Physics 9 (2), 190–194.

11

Application 2: shipping and offshore industry

Elzbieta M. Bitner-Gregersen
DNV, Høvik, Norway

Introduction

Rogue waves have attracted considerable attention in the scientific community, the media, and the shipping and offshore industries during the past two decades. The existence of exceptionally large waves—much higher, steeper, and more dangerous than those expected for a given sea state—has always been a part of maritime folklore. For centuries, sailors have reported giant waves, often described as "walls of water," appearing from "nowhere." In 1826, for example, the French scientist and naval officer Captain Dumont d'Urville reported having observed waves of up to 30 meters in height in the Indian Ocean grown as a wall of water. For a long time, such events were believed to be mostly anecdotal, but in recent decades new, more reliable measurements, the substantial increase in computational power, and advances in wave modeling have confirmed the existence of these abnormal waves.

The first real account of the rogue wave phenomenon in the scientific literature was given by Draper (1964), while Mallory (1974) provided initial discussions on the large and abnormal waves in the Agulhas current off the southeast coast of South Africa, the ocean region important for ship traffic. However, it is primarily during the last two decades that rogue waves have become the focus of wider attention from the scientific community and the marine industry, through larger research programs, meetings, workshops, and conferences all dedicated to the rogue wave phenomenon.

An important factor that has led to increased interest in rogue waves in the marine industry is the emergence of reliable observations of such waves provided by wave measurements from, for example, oil platforms. One of the first, and probably best-known, reliable recordings of a rogue wave is the so-called New Year wave, also referred to as the Draupner wave, that was recorded at the Draupner platform in the North Sea on January 1, 1995 (Haver and Andersen, 2000). Whereas

Science and Engineering of Freak Waves.
DOI: https://doi.org/10.1016/B978-0-323-91736-0.00010-9

previous recordings of extremely large waves were often regarded as measurement errors, the validity of the Draupner wave was confirmed by damage to equipment due to the wave hitting a platform deck.

The motivation for investigating and understanding rogue waves by the marine industry is obvious. Large waves represent a major danger to ships, offshore and marine installations, as investigations of ship and offshore structures' accidents have shown. It is believed that 22 large vessels were lost or severely damaged, with 542 human fatalities, between 1969 and 1994 due to bad weather (Kharif et al., 2009). Between 1995 and 1999, a total of about 650 ship accidents were reported due to bad weather (Guedes Soares et al., 2001; Toffoli et al., 2005). Public and media awareness of the dangers associated with extreme waves has also been increasing in recent years, largely due to accidents causing major environmental disasters with pollution of large coastal areas (e.g., oil tankers Erika in 1999 and Prestige in 2002) as well as incidents of rogue waves hitting passenger ships (e.g., Queen Elisabeth II in 1995; Caledonia Star and Bremen in 2000; Explorer, Voyager, and Norwegian Dawn in 2005; Louis Majesty in 2010; MS Marco Polo in 2014; and Quark Expeditions' World Explorer in 2022), some of which resulted in passenger and crewmember fatalities (e.g., Louis Majesty, MS Marco Polo, and Quark Expeditions' World Explorer accidents). Recent hurricanes in the Gulf of Mexico have also shown how dangerous extreme sea states can be for marine structures. During Hurricane Katrina, 30 oil platforms were damaged or destroyed, nine refineries were closed, and oil production in the region was decimated. The wave which struck the mobile offshore unit COSL Innovator in the North Sea on December 30, 2015, resulted in injuries of four crewmembers and one fatality. Such accidents have highlighted the need for improvements to reduce the risk of wave-related incidents.

In the future, the significance of severe sea state conditions for ship traffic may even grow in some ocean regions because of the expected increase in the frequency and severity of extreme weather events associated with global warming (Bitner-Gregersen and Toffoli, 2015; Gramstad et al., 2017; Bitner-Gregersen et al., 2018). Therefore taking rogue waves into account in the design and operation of marine structures may become an important part of the adaptation to climate change.

This chapter aims at providing the reader insight into the proactive role played by the shipping and offshore industries in investigations of these abnormal waves and presenting the status of implementation of rogue waves in marine industry standards. It discusses challenges associated with developing design criteria for these waves. Examples illustrating the impact of rogue wave effects on design loads are shown. Finally, future research activities that are still needed to support the possible systematic inclusion of rogue waves into ship and offshore structure design and marine operations are discussed.

The reader of this chapter is directed also to the DNV Position Paper on Rogue Waves (Bitner-Gregersen and Gramstad, 2015) and the DNV Feature Article "Rethinking Rogue Waves" (Bitner-Gregersen, 2017) which the present chapter summarized.

Activities on rogue waves in the marine industry

General

The marine industry has always been actively investigating ocean waves, particularly extreme waves, important for the design of marine structures; therefore rogue waves have also got attention. In the 1990s, the rogue wave phenomenon was still relatively poorly understood. However, literature reviews, participation in international conferences and workshops, and affiliations of the shipping and offshore community with leading researchers in the wave field contributed to the initiation of research on rogue waves in the marine industry in the end-1990s.

The recording of the New Year wave (known as the Draupner wave) at the Draupner platform in the North Sea on January 1, 1995 (Haver and Andersen, 2000), played an important role in this development. Although the damage to the equipment on the Draupner platform deck was minor, the validity of the Draupner wave was confirmed.

The marine industry activities on rogue waves presented herein are limited to European international and national research projects in which the Norwegian Classification Society Det Norske Veritas (DNV) participated (Bitner-Gregersen and Gramstad, 2015). It should be mentioned, however, that in the two largest projects the International Association of Classification Societies (IACS) was an observer. Oil companies participating in these projects have been members of the International Association of Oil & Gas Producers (IOGP).

The international projects in which DNV participated were partially, or fully, funded by the European Commission (EC) and the marine industry (Joint Industry Projects, JIPs). They, together with more recently two national Norwegian projects, are listed below.

EC MAXWAVE project

The idea for initiating the MaxWave (Rogue Waves—Forecast and Impact on Marine Structures) project arose during the 1998 COST 714 conference in Paris on directional wave spectra following an initiative from Prof. Douglas Faulkner. Ship accidents associated with bad weather were the main inspiration for starting the project.

The project was funded within the EC Fifth Framework Programme in the period 2000–03 and coordinated by GKSS Research Centre in Germany. It included four European Meteorological Offices (GKSS Research Centre, Norwegian Meteorological Institute, Meteo France, and UK Meteorological Office), three universities (Catholic University of Leuven, KU Leuven, in Belgium, Instituto Superior Técnico, IST, in Portugal, and Technical University of Berlin, TUB), two large research organizations (Institute of Hydro-Engineering of the Polish Academy of Sciences and German Aerospace Centre), one small/medium-sized enterprise (Ocean Waves, Germany), and one classification society (DNV, Norway). A project Senior Advisory Panel (SAP) functioned as an entrance to international bodies, and the project Expert Board brought knowledge from outside the project to the consortium. IACS was an observer in SAP.

The MaxWave project aimed to provide a better understanding of the physical and statistical properties of rogue waves, to develop risk maps and warning criteria for such extreme events, to investigate their impact on current design procedures, and to carry out a socioeconomic assessment of possible revisions of the rules (Rosenthal and Lehner, 2008). MaxWave made a significant contribution to our understanding of rogue waves. However, due to the complexity of the topic several important questions remained to be answered.

EC Network SEAMOCS

The need for further investigations of extreme weather events was also recognized by the EC, which fully funded the Marie Curie Network, SEAMOCS (Applied Stochastic Models for Ocean Engineering, Climate, and Safe Transportation), within the EC Sixth Framework Programme. This was a 4-year (2005–09) research training and mobility network coordinated by the University of Lund, Sweden. The SEAMOCS Network partners included Lund University, University of Sheffield in United Kingdom, University Paul Sabatier in Toulouse, France, KU Leuven in Belgium, Chalmers University of Technology in Sweden, Tallinn University of Technology in Estonia, Royal Netherlands Meteorological Institute, Swedish Meteorological and Hydrological Institute, and DNV in Norway. The overall goal of the network was to increase marine safety and to reduce the capital and operational costs of sea transport and major offshore installations.

DNV hosted a post-doc from KU Leuven who brought to DNV a nonlinear wave code based on the higher-order spectral method (HOSM) that was originally developed by the University of Turin, Italy.

EC EXTREME SEAS project

Investigations in the MaxWave project and the SEAMOCS Network inspired DNV to propose a new EC project—EXTREME SEAS (Design for Ship Safety in Extreme Seas) (Bitner-Gregersen, 2013). It was clear to DNV that several issues needed further investigation: the generation of rogue waves was not fully understood, agreement about the probability of occurrence of rogue waves had not been reached, and systematic investigations on the impact of these abnormal waves on current design procedures were lacking.

EXTREME SEAS was a collaborative project funded by the EC Seventh Framework Programme (2009–13). The project consortium consisted of two shipyards (MEYER WERFT (MW) from Germany and Estaleiros Navais Viana de Castelo from Portugal, ENVC), two classification societies (DNV and Germanischer Lloyd AG, GL), one model basin (CEHIPAR in Madrid, Spain), one provider of meteorological services (Norwegian Meteorological Institute), one research institute (Institute of Applied Physics of the Russian Academy of Sciences, RAS), and four universities (University of Turin in Italy, Instituto Superior Técnico in Portugal, and University of Duisburg-Essen and Technische Universität Berlin (TUB) in Germany). TUB operated a model tank, which was also used in the project. EXTREME SEAS was coordinated by DNV, Norway.

An EXTREME SEAS Senior Advisory Panel (SAP) was an observer in the project and supported the dissemination and exploitation of the project results. It consisted of IACS, Joint Committee of Marine Meteorology of WMO (JCOMM), Color Line, and Carnival. MW, through their role as Chairman of the Technical Committee of EUROYARDS, which represents six major shipyards in five European countries, was the exploitation manager of the project.

The overall objective of EXTREME SEAS was to provide the technology and methodology necessary for safe ship design in extreme sea conditions.

The investigations carried out in EXTREME SEAS contributed significantly to improving our understanding of rogue waves, as well as their impact on ship structures.

JIP CresT/ShortCresT CresT ShOrT

Hurricanes Ivan, Rita, and Katrina in the Gulf of Mexico demonstrated the dangers of extreme sea conditions for all types of offshore structures and drew considerable attention to the danger of extreme waves in the offshore industry. This resulted in the initiation of the 2-year JIP project Cooperative Research on Extreme Seas and their impacT (CresT) in 2008 (https://www.marin.nl/en/jips/crest9) which continued as the ShorTCresT (Effects of ShorTCrestness on wave impacT) JIP project. Both projects were coordinated by MARIN in the Netherlands and funded by oil companies, offshore engineering companies, and classification societies. The project partners included SHELL, Oceanweather Inc., Forristall Ocean Engineering Inc. from United States, Ocean Wave Engineering, Imperial College from United Kingdom, and DNV from Norway.

The CresT project studied unidirectional waves, whereas the ShorTCresT project focused on directionally spread seas. Model tests were carried out in the

MARIN and Imperial College wave basins. A generic tension-leg platform (TLP) was defined and chosen as a case study in both projects.

RCN ExWaCli ExWaCli

Requests from different end users within the marine sector regarding the implications climate change may have on metocean conditions, and consequently for the marine industries, were the reason for proposing a national research project Extreme Waves and Climate change: Accounting for uncertainties in design of marine structures (ExWaCli). The ExWaCli (2013–16) project was partially funded by the Research Council of Norway and coordinated by DNV, Norway (Bitner-Gregersen et al., 2018). The Norwegian Meteorological Institute and the University of Oslo (Department of Mathematics) were partners in the project.

The main objective of ExWaCli was to obtain an understanding of how wave conditions in the North Atlantic will be affected by climate change during the 21st century as well as specifying uncertainties associated with the predicted changes. Rogue waves in the future climate were also studied.

RCN ExWaMar ExWaMar

The work on the development of warning criteria for extreme and rogue waves was initiated already in the EC MaxWave project and continued in the EC EXTREME SEAS project, but the proposed criteria were not sufficiently general to capture all important aspects of rogue wave occurrences. This was a motivation for proposing a national research project EXtreme wave WArning criteria for MARine structures (ExWaMar). ExWaMar (2016–20) was funded partially by the Research Council of Norway and coordinated by DNV. The Norwegian Meteorological Institute and the University of Oslo (Department of Mathematics) were partners in the project.

The main objective of ExWaMar was to develop improved warning criteria that indicate increased risk of extreme or rogue waves, relevant for shipping and offshore industries. The project aimed also, among others, at the development of a DNV HOSM code which includes, apart from water surface elevation, also wave kinematics through the water column. A paper summarizing the project results is under prepartion.

Probability of occurrence of rogue waves in the context of design

The traditional format of classification societies' rules has been mainly prescriptive, without a transparent link to an overall safety objective. The International Maritime Organization (IMO, 1997, 2001) has developed guidelines for the use of the Formal Safety Assessment (FSA) methodology in rule development, which provide risk-based goal-oriented regulations that are well balanced with respect to acceptable risk levels and economic considerations. Risk is defined in FSA as a product of the probability of occurrence of a hazard multiplied by its consequence estimated usually as a monetary value. In the case of a rogue wave, its occurrence will represent a hazard. FSA is widely used in the offshore industry (ISO 2394, 1998) and has also started to be used within the shipping industry (DNV,1992, IACS, 2001, 2010).

In FSA, the probability of occurrence of a hazard during the lifetime (return period) of a structure, for example, 25 years or 100 years, is required. The frequency of occurrence of rogue waves was unknown for many years because of limited knowledge about rogue waves, too few measurements of rogue waves, and lack of consensus about a definition of rogue waves within the wave community. For design to provide probability of such hazard, frequency of occurrence of a rogue wave in a short-term stationary sea state, being typically of duration 20 min, 3 or 6 hours, and the probability of occurrence of this sea state within the lifetime of a structure need to be combined (Bitner-Gregersen and Hagen, 2004).

Owing to the last two decades' research efforts, the mechanisms generating rogue waves, their detailed dynamic properties, and metocean conditions in which they occur are now better understood, even though still debated in the literature. Despite these achievements, full consensus on the probability of the occurrence of rogue waves has yet to be achieved. Such consensus is essential to enable a systematic revision of offshore standards and classification societies' rules and is the reason why has not taken place (Bitner-Gregersen and Gramstad, 2015).

Rogue waves that exceed the simplified, commonly used criteria due to Haver (2000), $C_{max}/H_s > 1.25$ or/and $H_{max}/H_s > 2.0$, are expected to occur in a random wave field even within the standard linear and second-order description of random waves (Chapter 1). Taking into account the probability of occurrence of sea states in the North Sea scatter diagram and second-order distribution of Forristall (2000), Bitner-Gregersen and Hagen (2004) showed that in a random 20-minute sea state in a single point location, a wave satisfying $C_{max}/H_s > 1.3$ occurs with a probability of about $1.44 \cdot 10^{-3}$, corresponding to a return period of about eight days. However, such a wave is not necessarily dangerous. Whether or not a wave is dangerous is case-dependent, and is different, for example, for design of offshore platforms than for operation of small fishing vessels. Simultaneous high and steep waves typically give high loads on marine constructions. According to the second-order model combined with the North Sea scatter diagram, the probability of occurrence of a wave with $C_{max} > 4.5$ m and $C_{max}/H_s > 1.3$ is about 10^{-3}, whereas the probability of occurrence of a wave with $C_{max} > 8.5$ m and $C_{max}/H_s > 1.3$ is about 10^{-4}.

In design, it is common to use the most probable extreme wave (or the extreme mean value) in a wave record of a given duration. The expected extreme event depends on the duration of the wave record, and a longer time series allows the capture of more extreme waves. For example, the 6-hour sea state duration that is commonly used in design today is insufficient to capture the Draupner wave when the second-order wave model is applied (Bitner-Gregersen, 2003).

The second-order-based probabilities of occurrence of rogue waves are already accounted for in current design practice, if a second-order model is applied. An important question that remains is how much these probabilities increase if higher-order effects are included. Situations occur where significant deviations from second-order statistics have been observed, such as the cases of steep sea states and narrow directional wave spreading, crossing seas, and other effects (Bitner-Gregersen and Gramstad, 2015; Gramstad et al., 2018a). These situations responsible for the generation of rogue waves have been observed in theoretical and numerical wave models as well as laboratory experiments. Exactly how frequently such situations occur in the ocean is still a topic of investigation.

In the MaxWave project, several extreme waves, exceeding the wave height criterion $H_{max}/H_s > 2$, were detected in a global satellite SAR data set acquired over three weeks in the South Atlantic (Rosenthal and Lehner, 2008). Bitner-Gregersen and Toffoli (2012) analyzed 10-year hindcast data from the North Atlantic in the EXTREME SEAS project and showed that sea states for which modulational instability may be active occurred several times during the period the data covered.

In another EXTREME SEAS study, Bitner-Gregersen and Toffoli (2014) using 50-year hindcast data showed that crossing wave systems associated with increased occurrence of rogue waves occurred several times in the North Atlantic and the North and Norwegian Seas. These crossing sea states, which are characterized by two wave systems with approximately the same energies and peak frequencies, and a crossing angle of about $40^\circ - 50^\circ$ (see e.g. Cavaleri et al., 2012) were mainly observed in low and intermediate sea states, what later also was confirmed by Gramstad et al. (2017) and Vettor and Guedes Soares (2020). The third-order HOSM simulations carried out by Bitner-Gregersen and Toffoli (2014) showed increased rogue wave occurrence not only when the two wave systems had very narrow directional distributions, but also for broader directional spectra. Recently, Gramstad et al. (2018a) demonstrated by the third-order HOSM simulations that maximum sea surface kurtosis, an indicator of the occurrence of rogue waves, is expected not only for small crossing angles but also for relatively large angles, with a minimum around 90 degrees.

Investigations carried out in the ExWaMar project by Gramstad et al. (2018b) of 1-year in situ measurements from the central North Sea from three different sensors, a waverider buoy, a downlooking laser array, and a WaveRadar REX, showed that rogue waves occur on average about every 7–8 days in a single point location.

The above findings, apart from the satellite SAR data, refer to the occurrence of rogue waves at a single point location. Recent investigations of Bitner-Gregersen et al. (2020) by use of the third-order numerical HOSM simulations show that a much higher frequency of occurrence of extreme and rogue waves can be expected if spatial–temporal data are considered. Therefore to specify the probability of occurrence of rogue waves in a sea state use of stereo-video camera systems (Benetazzo et al., 2015), which may cover relatively large observation domains, up to O(100–1000) m^2, is important (Chapter 4).

Climate change and rogue waves

Global warming and extreme weather events reported in the last years have attracted a lot of attention not only in academia (Hemer et al., 2012; Aarnes et al., 2017; Takbash and Young, 2020) and media but also in the shipping as well as offshore and renewable energy industry (Bitner-Gregersen et al., 2013, 2018; Hagen et al., 2013). A central question for the marine industries has been: to what degree will climate change affect future ship traffic and design of ships as well as offshore structures and marine installations. There is also a concern in academia and the marine industries regarding the frequency of occurrence of rogue waves in the future climate. Climate change and rogue waves represent two different phenomena. While climate change is due to warning of the Earth surface compared to the preindustrial period, rogue waves are generated by different mechanisms and have always been and will be present in the ocean.

The observed climate changes include natural variability of climate and anthropogenic climate change which is due to human activities and results in the observed warming of the Earth's surface (IPCC, 2022). This also leads to changes in metocean conditions. Changes in wave and wind climate are expected to have the largest impact on marine structure design in comparison with other environmental phenomena since for most marine structures wave- and wind-induced loads are dominating.

Climate change in terms of increased storm activity (intensity, duration, and fetch) in some ocean areas, and changes of storm tracks, may lead to secondary effects such as increased frequency of occurrence of rogue waves (Bitner-Gregersen and Toffoli, 2015; Gramstad et al., 2017).

At present, there are large uncertainties associated with climate change projections (Aarnes et al., 2017; Bitner-Gregersen et al., 2018). They include large variability between the different climate models and across different ensemble members of the same model as well as between the analyzed emission scenarios. Also, adopted downscaling techniques (dynamical and statistical) impact projected changes of wave climate, which are location-dependent. These uncertainties will also affect predictions of sea states which may trigger the occurrence of rogue waves. Nevertheless, the investigations carried out in the ExWaCli project showed that, for example, in the northernmost location that was considered, outside the Norwegian coast of Finnmark, a significant increase, in the range of 110%–360%, in the number of rogue-prone crossing seas (wind sea and swell) has been observed for all six climate models used by the project (Gramstad et al., 2017). It should also be noticed that in the polar regions the receding ice coverage will obviously change the wave climate, and higher and steeper waves can be expected. This is strongly location-dependent.

Firm conclusions regarding future changes in metocean conditions and the frequency of occurrence of rogue waves in the future climate are difficult to reach today. The climate community is continuously improving climate change projections. The recent findings are summarized in the IPCC Sixth Assessment Report released in 2022, which is under evaluation by the marine industry.

Impact of rogue waves on loads and responses

Formal recognition of the existence of rogue waves has caused understandable concern about the impact such waves may have on ships and offshore structures. Several studies conducted so far demonstrated that these waves may have a significant effect on the loads and responses of ships and offshore structures. They may affect both global and local loads of ships and offshore structures and, consequently, their design. Below are some examples describing different proposed procedures for including rogue waves in load and response calculations and how their impact load and response prediction are given.

In the MaxWave project, Fonseca et al. (2001) proposed a simplified method using measured sea surface elevations containing abnormal wave events as input to the IST strip theory code, while Pastoor et al. (2003) demonstrated how such measured wave records can be used as input to the DNV 3D panel code. A method based on measured sea surface elevations containing abnormal wave events as input data to a wave–structure interaction code showed good agreement with model test data by TUB.

EXTREME SEAS was the first project to investigate the impact of rogue waves on ship structures more systematically, by using both model tests and numerical simulations. The case studies addressed by EXTREME SEAS included a container vessel, a passenger ship, an LNG carrier, and a product and chemical tanker. Ship behavior in extreme and rogue waves was investigated by the numerical sea-keeping codes enhanced/developed within the project: strip theory code, 3D panel code, and computational fluid dynamics (CFD). The numerical results were validated by model tests carried out in the TUB tank and the Spanish basin Canal de Experiencias Hidrodinamicas de El Pardo (CEHIPAR). It was demonstrated that rogue waves have a significant impact on wave-induced bending moments (sagging and hogging), slamming, and ship motions such as heave, pitch, and surge (Guo et al., 2013; Ley, 2013). The strip theory and 3D panel codes were not able to capture the effects of rogue waves satisfactorily. The CFD-based approach gave better predictions than the two other methods, even though still not completely satisfactory. At the TUB, breather solutions of the nonlinear Schrödinger equation (NLS) were successfully reproduced in a wave tank with the help of the University of Turin and, for the first time, used in sea-keeping tests (Clauss et al., 2012; Chapter 13), providing new perspectives on the methodology of examining ships and offshore structures in rogue waves.

The German shipyard Meyer Werft (MW), in collaboration with legacy GL Hamburg and the University of Duisburg-Essen, showed significant impact loads from rogue waves on the superstructure of a cruise ship built by MW (Fig. 11.1) and proposed redesign of the superstructure in EXTREME SEAS.

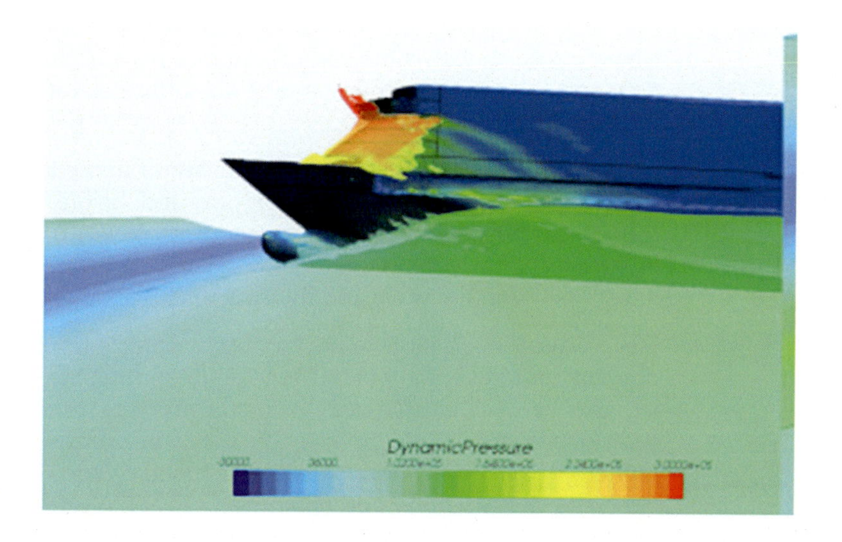

Figure 11.1 The cruise ship analyzed in EXTREME SEAS by MW.

It is crucial that the design of offshore structures ensures that platform decks are safe from green water events. In the CresT project, particular focus was on an extreme sea state that the platform Marco Polo in the Gulf of Mexico experienced during Hurricane Rita. This sea state was applied in an analysis of the behavior of a generic TLP system defined by the project. In the reliability assessment of the air gap for this generic TLP, the Forristall (2000) second-order wave model for long-crested (2D) and short-crested (3D) sea, with a correction factor for higher-order nonlinearities, was used by Bitner-Gregersen (2011) (Fig. 11.2). The instantaneous air gap (DNV, 2019) was determined from TLP offset combined with instantaneous wave crest elevation:

$$a(x,y,t) = a_0 - \left(\eta_{ise}(x,y,t) - z_{setdown}(x,y,t)\right) \tag{11.1}$$

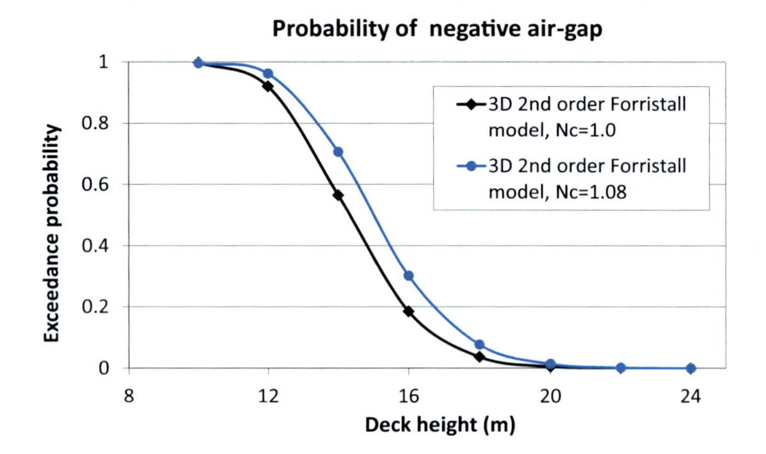

Figure 11.2 Probability of negative air gap $a(x, y, t) < 0$ for different platform deck heights and different degrees of nonlinearity of the sea surface introduced to the 3D Forristall (2000) model. 3-hour sea state duration.

where $z_{setdown}(x, y, t)$ is the vertical displacement of the structure at (x, y) position, and $\eta_{ise}(x, y, t)$ is the instantaneous surface elevation at the same horizontal position at (x, y) at time t.

Fig. 11.2 shows increase in the probability of exceedance of negative air gap $a(x, y, t) < 0$ as a function of the platform deck height when the correction factor due to nonlinear effects is increased from $Nc = 1.0$, referring to the second-order effects, to $Nc = 1.08$. Negative air gap $a(x, y, t) < 0$ means that there is impact between the wave and the structure.

Steep waves have a large impact on water particle kinematics, for which proper descriptions are very important for design and marine operations. The increasing use of CFD in ship and offshore structure assessment requires proper descriptions of very steep waves that can be obtained from nonlinear wave models. In DNV, a nonlinear wave code based on the HOSM is used in this process. The DNV HOSM code (Gramstad, 2017) in addition to wave properties on the water surface, which traditionally HOSM codes provide, allows for the prediction of wave kinematics in the water column.

Current design practice

There is potential safety, environmental, and economic advantage in utilizing the most recent knowledge about metocean description in the operation and design of marine structures. Waves represent the dominant environmental load for most marine structures. Therefore understanding, and possibly predicting, waves under various conditions is crucial with respect to the design and operation of ships, offshore structures, and marine installations. To achieve acceptance, a metocean description must be demonstrated to be robust and of adequate accuracy. As with most formal processes, updating codes and standards takes some time, and consequently, updates may lag behind the state of the art.

Classification societies' rules and offshore standards are dynamic and continuously updated to account for state-of-the-art knowledge. The occurrence of rogue waves poses two important questions for the shipping and offshore industries: *Should these waves be accounted for in design? If so, how best to account for them?*

The majority of ocean-going ships are currently designed for the North Atlantic wave environment, which is regarded as the most severe. Visual observations of waves collected from ships in normal service and summarized in the British Maritime Technology Global Wave Statistics (GWS) atlas (BMT British Maritime Technology, 1986, updated in the 1990s) are used for ship design and operations. The average wave climate of four ocean areas in the North Atlantic is recommended by IACS (2001) for ship design. It should be noticed that the North Atlantic scatter diagram is currently under revision by classification societies to be replaced by more accurate global wave data which are available today (de Hauteclocque et al., 2020).

Unlike ship structures, offshore structures normally operate at fixed locations and often represent a unique design. As a result, platform design and operational conditions need to be based on the location-specific metocean environment (NORSOK, 2017). Because of the limited availability of measurements and improved hindcasts, the latter have become increasingly relied upon in design and operations during the last decades.

Introduction by IMO Guidelines (IMO, 1997, IMO, 2001) for application of the FSA methodology in rule development allows the provision of risk-based goal-oriented regulations. Structural reliability analysis (SRA) (Madsen et al., 1986) is used in the development of marine industry standards. To determine design loads, different limit state categories and scenarios are used (DNV, 1992).

In the design process, international standards are followed to calculate ship structural strength and ship stability during extreme events, with a return period of 25 years; the ultimate limit state (ULS) check corresponds to the maximum load-carrying resistance. Offshore structures follow a different approach to that of ship structures and are designed for the 100-year return period (ULS). In addition, the Norwegian offshore standards (NORSOK, 2017) require that there must be enough room for the wave crest to pass beneath the platform deck to ensure that a 10,000-year wave load does not endanger the structural integrity (Accidental Limit State, ALS).

Traditionally, the shipping industry has used linear irregular waves as input to numerical codes for calculations of structural loads and responses, while nonlinear second-order irregular waves are applied currently by the offshore industry when assessing loads and responses of offshore structures. Neither linear nor second-order wave models are able to realistically describe very steep waves, and both models fail to correctly predict abnormal events such as rogue waves. The introduction of nonlinear waves beyond the second order as input to the wave–structure interaction codes is necessary in this case. The increasing use of CFD in ship and offshore structure analysis also requires proper descriptions of very steep waves that can be obtained from nonlinear wave models.

The main concern of the shipping and offshore industries is how the safety level currently used in design will be affected by the presence of rogue waves. To investigate it, the IACS Common Structural Rules for Tankers (CSR) (IACS, 2010), which address hull girder collapse of tankers in sagging conditions, were considered in EXTREME SEAS (Bitner-Gregersen et al., 2015). The analyses included the same ships as considered by IACS (2010): Suezmax, Product Tanker, VLCC 1, VLCC 2, and Aframax. To account for rogue waves, a simplified approach was applied in which the annual extreme vertical wave bending moment was scaled by a constant factor of between 5% and 30%, reflecting the findings of Guo et al. (2013). The annual probability of failure as a function of deck area (total cross-sectional area of deck plate and stiffeners) for the Aframax test tanker is shown in Fig. 11.3. The tanker length is 234 m, its breadth is 42 m, and its depth is 21 m. The deck area equal to 1 refers to the initial input design as adopted in the CSR.

Fig. 11.3 demonstrates that the probability of failure of Aframax increases for an increase in the extreme bending moment. If the extreme bending moment is increased by 10%, the corresponding increase in the deck area (steel weight of the deck in the midship region) is also around 10%, needed to maintain the reliability level. The same overall trend for all the tankers analyzed was observed.

The increased costs of marine structures when design accounts for rogue waves are still not sufficiently investigated but can be kept low by the introduction of

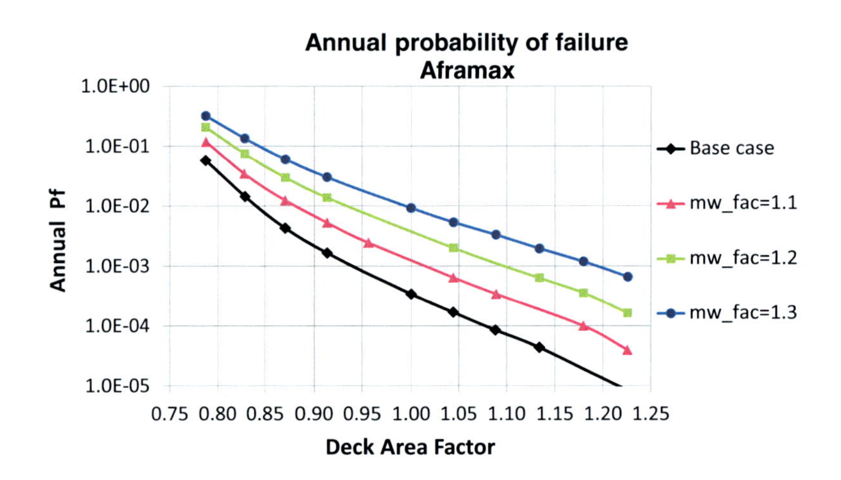

Figure 11.3 Annual probability of failure, Aframax. Base case, effects of rogue waves.

innovative design as demonstrated in the EC EXTREME SEAS project by the Portuguese shipyard, ENVC.

A systematic revision of classification society rules and offshore standards has not taken place due to a lack of consensus about the probability of occurrence of rogue waves in the ocean, which is mandatory (Bitner-Gregersen & Gramstad, 2015). However, research findings carried out so far have led to some updates of rules and offshore standards.

The investigations of the German shipyard Meyer Werft, in collaboration with legacy GL Hamburg and the University of Duisburg-Essen, carried out in the aforementioned EXTREME SEAS project, resulted in updating of DNV rules for the design of superstructures of passenger ships to account for rogue wave effects.

In 2015, DNV established an interface between the DNV nonlinear wave HOSM solver and the DNV wave–structure interaction 3D panel code, WASIM. This enabled the sea-keeping code to be run with more realistic nonlinear waves as input. A subsequent internal study extended this work and developed an interface between HOSM and the CFD code Star-CCM+, providing nonlinear wave kinematics in the CFD domain covering the structure. The interface was validated through a case study of a semisubmersible, and the effects of using linear and nonlinear waves as input for predictions of loads and responses were investigated. It was shown that when using more realistic nonlinear wave input, a smaller CFD domain may be used, saving precious and expensive computational time.

Rogue waves are implemented in DNV Recommended Practice RP-C205 (2019) while the DNV HOSM solver is a part of the DNV SESAM Software System. Notice that HOSM can be used in direct calculations of loads and responses (time-domain simulations) which the DNV rules allow to be used in assessment of performance of ship structures.

Furthermore, following the lead set by Equinor a simple requirement that accounts for rogue waves, when designing the height of a platform deck, was introduced recently in the revised version of the Norwegian Standard (NORSOK, 2017). Indeed, NORSOK (2017) recommends a 10% increase in estimated extreme crest height compared to second-order point statistics to account for nonlinear waves and spatial effects when designing the height of a platform deck.

Conclusions

The marine industry has methods and tools, even though some may need to be enhanced, which allow to investigate effects of rogue waves on loads and responses of marine structures as well as to account for rogue waves in design. The second- and higher-order wave codes, received from academia, developed in-house (e.g., DNV HOSM code) or open-access codes (e.g., the Nantes group HOSM code), are available to the marine industry.

Knowledge about the probability of rogue waves in the ocean is mandatory to start systematic revision of the classification societies' rules and offshore standards. Investigations on climate change which may lead to an increased frequency of occurrence of rogue waves are still ongoing.

It should be underlined that not all wave loads and responses of marine structures will be equally sensitive to the nonlinearity beyond the linear and second order. The current design practice includes apart from adopted models also safety factors. Therefore further investigation of rogue wave effects on loads and responses of marine structures is necessary to be able to evaluate the existing margins in the current design.

Retaining the current safety level in rules and standards during the possible systematic implementation of rogue waves is crucial. Distinction will need to be made between the existing and future marine structures, and different types of marine structures when accounting for these extreme waves in design.

Collaboration of the marine industry with the academia should continue to reach firm conclusions regarding possible systematic implementation of rogue waves in the marine industry standards.

Acknowledgments

DNV acknowledges the EC and the Research Council of Norway for funding research on rogue waves. The author expresses her thanks to Dr Odin Gramstad for his comments.

References

Aarnes, O.J., Reistad, M., Breivik, O., Bitner-Gregersen, E., Eide, I.L., Gramstad, O., et al., 2017. Projected changes in significant wave height towards the end of the 21st century: Northeast Atlantic. Journal of Geophysical Research: Oceans 122, 3394–3403.

Benetazzo, A., Barrariol, F., Bergamasco, F., Torsello, A., Carniel, S., Sclavo, M., 2015. Observation of extreme sea waves in a space–time ensemble. American Meteorological Society 45, 2261–2275.

Bitner-Gregersen, E.M., 2003. Sea state duration and probability of occurrence of a freak wave. In: Proceedings of the Twenty-Second International Conference on Offshore Mechanics and Arctic Engineering Conference; Cancun, Mexico, 8–13 June 2003.

Bitner-Gregersen, E.M., Hagen, Ø., 2004. Freak wave events within the 2nd order wave model. In: Proceedings of the Twenty-Third International Conference on Offshore Mechanics and Arctic Engineering; Vancouver, Canada, 20–25 June 2004.

Bitner-Gregersen, E.M., 2011. Reliability assessment of TLP air-gap in nonlinear waves. In: Proceedings of the Thirtieth International Conference on Ocean, Offshore and Arctic Engineering; Rotterdam, The Netherlands, 19–24 June 2011.

Bitner-Gregersen, E.M., Toffoli, A., 2012. On the probability of occurrence of rogue waves. Natural Hazards and Earth System Sciences 12, 751–762.

Bitner-Gregersen, E.M., Lars, I.E., Hørte, T., Skjong, R., 2013. Ship and Offshore Structure Design in Climate Change Perspective. Monograph, Springer Brief in Climate Studies, 28. ISBN 978-3-642-34137-3.

Bitner-Gregersen, E.M., Toffoli, A., 2014. Occurrence of rogue sea states and consequences for marine structures. Ocean Dynamics 64, 1457–1468.
Bitner-Gregersen, E.M., 2013. EXTREME SEAS Final Report. Grant SCP8-GA-2009–234175. http://cordis.europa.eu/result/rcn/55382_en.html.
Bitner-Gregersen, E.M., Eide, L.I., Hørte, T., Vanem, E., 2015. Impact of Climate Change and Extreme waves on Tanker Design. SNAME Transactions SNAME-SMC-2014-T43.
Bitner-Gregersen, E.M., Toffoli A., 2015. Wave steepness and rogue waves in the changing climate in the North Atlantic. In: Proceedings of the Thirty-Fourth International Conference on Ocean, Offshore and Arctic Engineering, St. John's, Canada, 31 May-5 June 2015.
Bitner-Gregersen, E.M., Gramstad, O., 2015. DNV Position Paper on Rogue Waves. https://www.dnv.com/publications/rogue-waves-60134.
Bitner-Gregersen, E.M., 2017. DNV Feature Article "Rethinking Rogue Waves," file:///C:/Users/Bruker/Downloads/DNV_GL_Feature_-_Rogue_waves_-_June_2017%20(1).pdf.
Bitner-Gregersen, E.M., Vanem, E., Gramstad, O., Hørte, T., Aarnes, O.J., Reistad, M., et al., 2018. Climate change and safe design of ship structures. Ocean Engineering 149, 226–237.
Bitner-Gregersen, E.M., Gramstad, O., Magnusson, A.K., Malila, M., 2020. Extreme wave events and sampling variability. Ocean Dynamics 71, 81–95.
BMT (British Maritime Technology), 1986. Primary Contributors: Hogben N, Da Cunha LF, Oliver HN. Global Wave Statistics. Unwin Brothers Limited, London, England.
Cavaleri, L., Bertotti, L., Torrisi, L., Bitner-Gregersen, E., Serio, M., Onorato, M., 2012. Rogue waves in crossing seas: The Louis Majesty accident. Journal of Geophysical Research 117.
Clauss, G., Klein, M., Dudek, M., Onorato, M., 2012. Application of breather solutions for the investigation of wave/structure interaction in high steep waves. In: Proceedings of the Thirty-First International Conference on Ocean, Offshore and Arctic Engineering, Rio de Janeiro, Brazil, 1–6 June 2012.
Draper, L., 1964. 'Freak' Ocean Waves. Oceanus 10, 13–15.
DNV, 1992. Structural Reliability Analysis of Marine Structures. Classification Note 30.6. July.
DNV, 2019. Environmental Conditions and Environmental Loads. DNV Recommended Practice, DNV-RP-C205. Høvik, Norway.
Fonseca, N., Guedes Soares, C., Pascoal, R., 2001. Prediction of ship dynamic loads in heavy weather. In: Proceedings of the Design and Operation for Abnormal Conditions II, London. Royal Institution of Naval Architects.
Forristall, G.Z., 2000. Wave crest distributions: Observations and second-order theory. Journal of Physical Oceanography 30 (8), 1931–1943.
Gramstad, O., 2017. Implementation of Higher-Order Spectral Method for Nonlinear Wave Simulations with Calculation of Water Particle Kinematics. DNV Technical Report. Høvik, Norway.
Gramstad, O., Bitner-Gregersen, E.M., Vanem, E., 2017. Projected changes in the occurrence of extreme and rogue waves in future climate in the North-Atlantic. In: Proceedings of the Thirty-Sixth International Conference on Ocean, Offshore and Arctic Engineering, Trondheim, 23–30 June 2017.
Gramstad, O., Bitner-Gregersen, E.M., Trulsen, K., Nieto Borge, J.C., 2018a. Modulational instability and rogue waves in crossing sea states. Journal of Physical Oceanography 48, 1317–1331.
Gramstad, O., Bitner-Gregersen, E.M., Breivik, Ø., Magnusson, A.K., Reistad, M., Aarnes, O.J., 2018b. Analysis of Rogue Waves in North Sea in-situ wave data. In: Proceedings of the Thirty-Seventh International Conference on Ocean, Offshore and Arctic Engineering, Madrid, Spain, 17–22 June 2018.
Guedes Soares, C., Bitner-Gregersen, E.M., Antao, P., 2001. Analysis of the frequency of ship accidents under severe North Atlantic Weather Conditions. Proceedings of Design and Operation for Abnormal Conditions II. London. Royal Institution of Naval Architects (RINA).

Guo, B.J., Bitner-Gregersen, E.M., Sun, H., Helmers, J., 2013. Prediction of ship response statistics in extreme seas using model test data and numerical simulation based on the Rankine Panel Method. In: Proceedings of the International Conference on Ocean, Offshore, and Arctic Engineering, Nantes, France, 9–14 June 2013.

de Hauteclocque, G., Zhu, T., Johnson, M., Austefjord, H., Bitner-Gregersen, E., 2020. Assessment of Global Wave Dataset for Long Term Response of Ships. In: Proceedings of the Thirty-Ninth International Conference on Ocean, Offshore and Arctic Engineering, Fort Lauderdale, FL, USA, 3–8 August 2020.

Hagen, Ø., Garrè, L., Friis-Hansen, P., 2013. DNV ADAPT framework for risk-based adaptation: a test case for the offshore industry. In: Proceedings of the ICOSSAR Conference New York.

Haver, S., 2000. Evidences of the existence of freak waves. Proceedings of the Rogue Waves; 129–140, Ifremer, Brest, France.

Haver, S., Andersen O.J., 2000. Freak Waves: Rare Realizations of a Typical Population or Typical Realizations of a Rare Population. In: Proceedings of the ISOPE Conference, Seattle, USA, 28 May- 2 June 2000.

Hemer, M.A., Wang, X.L., Weiss, R., Swail, V.R., 2012. Advancing wind-waves climate science: The COWCLIP project. Bulletin of the American Meteorological Society 93 (2012), 791–796.

IACS, 2001. International Association of Classification Societies (IACS) Rec. 34. Standard Wave Data. Rev.1 (London: IACS).

IACS, 2010. Common Structural Rules for Double Hull Oil Tankers with Length 150 Metres and above. Rules for Classification of Ships, 8, 1, July.

IMO, 1997. Interim guidelines for the application of formal safety assessment (FSA) to the IMO rule making process. In: Maritime Safety Committee, 68th Session, June; and Marine Environment Protection Committee, 40th Session, Sept. 1997.

IMO, 2001. Guidelines for formal safety assessment for the IMO rule making process. In: IMO/ Marine Safety Committee 74/WP.19.

IPCC, 2022. The Sixth Assessment Report: Climate Change (AR6) the Intergovernmental Panel on Climate Change. Cambridge University Press, Cambridge, UK and New York, USA.

ISO, 2394, 1998. General Principles on Reliability for Structures.

Kharif, C., Pelinovsky, E., Slunyaev, A., 2009. Rogue Waves in the Ocean. Springer.

Ley, J., 2013. Report with Validation of the New Time Domain Seakeeping Code by Systematic Comparisons with Experimental Data Obtained in the Project EXTREME SEAS report. University of Duisburg-Essen.

Madsen, H.O., Krenk, S., Lind, N.C., 1986. Methods of Structural Safety. Prentice-Hall, Enlewood Cliffs, NJ 07632.

Mallory, J.K., 1974. Abnormal Waves in the South-East Coast of South Africa. International Hydrogen Review 51, 89–129.

NORSOK, 2017. Standard N-003: Action and action effects. Rev. Jan.

Pastoor, W., Helmers, J.B., Bitner-Gregersen, E.M., 2003. Time simulation of ocean-going structures in extreme waves. In: Proceedings of the OMAE Conference, Cancun, Mexico, 8–13 June 2003.

Rosenthal, W., Lehner, S., 2008. Rogue waves: results of the maxwave project. Journal of Offshore Mechanics and Arctic Engineering 130, 021006.

Takbash, A., Young, I.R., 2020. Long-term and seasonal trends in global wave height extremes derived from ERA-5 reanalysis data. Journal of Marine Science and Engineering 8, 1015.

Toffoli, A., Lefèvre, J.M., Bitner-Gregersen, E.M., Monbaliu, J., 2005. Towards the identification of warning criteria: analysis of a ship accident database. Applied Ocean Research 27, 281–291.

Vettor, R., Guedes, Soares, C., 2020. A global view on bimodal wave spectra and crossing seas from ERA-interim. Ocean Engineering 210, 107439.

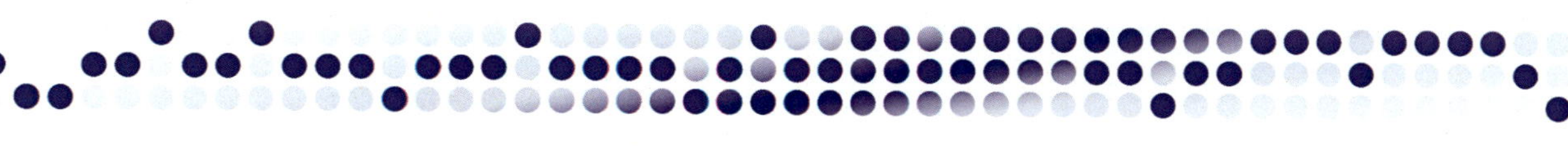

Application 3: extreme waves and coastal structures

Hiroaki Kashima[1] *and Nobuhito Mori*[2]

[1]Coastal and Ocean Development Group, Port and Airport Research Institute, Yokosuka, Kanagawa, Japan [2]Disaster Prevention Research Institute, Kyoto University, Japan

Introduction

As waves propagate from deep to shallow water, their amplitude steepens, and their wavelength shortens, while the frequency is constant for stationary conditions (Goda, 2000). The nonlinear stability of the wave trains due to the four wave—wave interactions becomes weaker in shallower water, although the second-order wave—wave interactions (bound mode effects) become dominant. A wave train is modulationally unstable in deep water at the critical depth $kh > 1.363$, while the steep-crested waves become modulationally stable in shallow water at the critical depth $kh < 1.363$ (Benjamin, 1967; Whitham, 1974), k being the wave number that corresponds to the water depth h. When the water depth becomes shallower than the critical depth $kh < 1.363$, the energy loss induced by the modulational instability may lead to reduce extreme wave occurrence (Mori and Janssen, 2006).

More recently, nonlinear behavior in shallow water after changes in water depth has been clarified in model experiments and numerical simulations as reported by Zeng and Trulsen (2012), Trulsen et al. (2012), Kashima et al. (2013, 2014), Kashima and Mori (2019), and Zhang et al. (2019). Many previous studies assumed a flat-bottom and quasi-stationary conditions for given water depth because the timescale of four-wave interactions is quite long as $O(\varepsilon^{-3}) - O(\varepsilon^{-4})$ in comparison with the second-order nonlinearity $O(\varepsilon^{-2})$ where ε is characteristic wave steepness. However, a wave transition sometimes occurs quickly from deep to shallow water on a slope. Therefore it is interesting to examine the transient behavior of the extreme waves, including the wave group enhanced in the deep water propagating to shallow water.

Science and Engineering of Freak Waves.
DOI: https://doi.org/10.1016/B978-0-323-91736-0.00012-2

Some extreme waves have been observed near the coastal area where the shallow water effects may become important (Janssen and Onorato, 2007). Although the depth-limited condition restricts the extreme wave heights in deep water, changes in wave height distribution depend on bottom topography and other factors (Mendez et al., 2004). The deep comprehension of the shallow water effects related to the extreme wave occurrence may be attributed to the proper design of the coastal and harbor structures. It is well known that the breakwaters are designed to provide good anchorage, moorage, and the safe navigation of ships and to promote the calmness of the harbor for wind waves, including extreme waves. The wave pressures on the front of the vertical wall of the breakwater are expressed by the wave pressure formula proposed by Goda (1973). The maximum wave height in a random wave train is used in this formula. It is based on the principle that a breakwater should be designed to be safe against a single wave with the largest pressure among random wave trains (Itoh et al., 1966).

The maximum wave height in engineering design is widely simply assumed as 1.8 times the significant wave height of the progressive wave train according to the Rayleigh theory, taking into account the performance of many prototype breakwaters as well as concerning the accuracy of the wave pressure estimation. Although the maximum wave height based on the significant wave height is constant, the real maximum wave height can be varied as a part of the dynamic and stochastic process of random waves and depth-limited effects. Therefore, in the point of the breakwater design from deeper to shallower water, it is necessary to evaluate two different effects, the statistical variability of maximum wave height and nonlinearity and depth-limited effects, on the wave pressures acting on the breakwater under the extreme wave conditions, appropriately. It is also interesting how deep water generated extreme waves by the four-wave interactions sustained in intermediate to shallow water conditions.

Methodology

Experimental conditions

The model experiments were performed in a two-dimensional wave tank located at the Port and Airport Research Institute in Japan. The wave tank is 35 m long, 0.6 m wide, and 1.5 m deep. The wave maker is of the piston type with the computer-controlled absorption installed at one end of the wave tank and controlled by a computer. The bottom model was selected to understand the nonlinear properties related to the extreme wave occurrence and wave pressure characteristics on the vertical breakwater in intermediate water. The sketch of the bottom model is shown in Fig. 12.1. The boxes show the wave gauges used to measure the water surface elevation in the wave tank. This bottom model was made of wood (rigid, smooth, and impermeable). The model

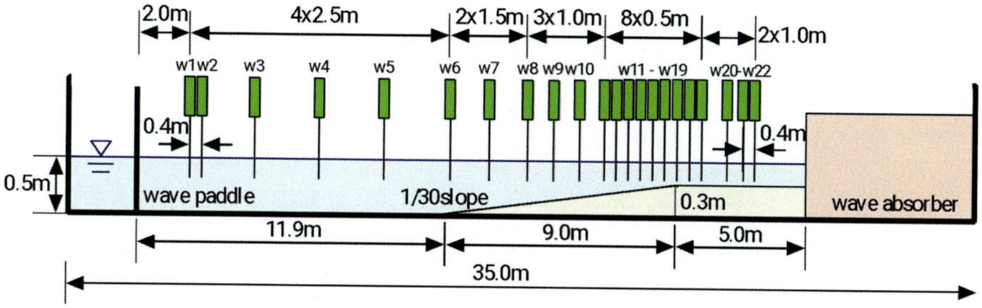

Figure 12.1 Schematic view of bottom configuration and position of wave gauges.

Table 12.1 Input wave parameters in model experiment.

$k_c h_i$	ε	Q_p	BFI
2.077	0.066	4.70	1.45

has a fixed impermeable 1/30 slope bottom installed at the toe 11.9 m from the wave maker and a flat mound of which the constant water depth is 0.2 m.

The input wave spectra as the signal transmitted to the wave maker were given by the JONSWAP spectra (Hasselmann et al., 1973). The initial water surface elevation was given as the linear combinations of sinusoidal waves through the wave spectra characterized by a significant wave period of 1.0 second. The values of the spectral parameters at the wave maker, the dimensionless water depth $k_c h_i$, the wave steepness $\varepsilon = k_c {m_0}^{1/2}$, Goda's spectral bandwidth parameter Q_p (Goda, 1970), and BFI are summarized in Table 12.1. Herein, h_i is the water depth in front of the wave maker, k_c is defined as the carrier linear wave number based on the significant wave period of 1.0 second and h_i, and m_0 is the variance of the water surface elevation. The BFI is expressed by the wave steepness and the spectral bandwidth and is given as $\varepsilon {Q_p}^2$ (Janssen, 2003).

Data analysis

Wave measurements were conducted along the center axis of the wave tank by using 22 capacitance-type wave gauges (Fig. 12.1). The water surface elevation was recorded with a sampling frequency of 20 Hz. Note that a large number of waves are of fundamental importance for the convergence of the tail of the probability density function for wave heights. Hence, the long-time model experiments are important to verify the effect of the number of waves from the point of view of understanding the maximum wave height correctly. Therefore, to guarantee the sensitivity of the wave statistics, such as the frequency distribution of the maximum wave height, ten measurement sets for the random water surface from a given spectrum were performed by using the ten sets of the random phases related to the randomness of wave trains. The total number of individual wave heights recorded at each wave gauge was about 10,000 waves. The duration of each measurement set was 25 minutes. For the present tests, we have removed the first 5 minutes of the records for each measurement set. Thus there are 20 minutes left which we will use for our analysis.

The individual wave heights recorded at each wave gauge were obtained by applying the zero cross-down method to the measured water surface elevation. The characteristics of the maximum wave height H_{max} are affected by the number of wave heights as described in Mori et al. (2007). Goda (2000) proposed that the maximum wave height is defined as the highest 1/250th wave height, $H_{1/250}$. Note that this definition yields the approximate relation of $H_{1/250} = 1.74\ H_{1/3}$ outside the surf zone according to the Rayleigh theory. It should be mentioned that an extreme wave is defined as one having a maximum wave height exceedingly not twice in Mori and Janssen (2006) and Mori et al. (2007) but 1.74 times the significant wave height from the point of view of the deviation from the Rayleigh theory. In the analysis, the number of the wave height for one wave train is defined as $N = 250$, and the maximum wave height in this study means the maximum value out of the 250 wave heights, approximately. Hence, we get 40 wave statistics from about 10,000 waves at each wave gauge. The skewness μ_3 and kurtosis μ_4 of the water surface elevation are defined as, respectively:

$$\mu_3 = \frac{1}{\eta_{rms}^3} \cdot \frac{1}{n} \sum_{j=1}^{n} \left(\eta_j - \eta_m \right)^3 \tag{12.1}$$

$$\mu_4 = \frac{1}{\eta_{rms}^4} \cdot \frac{1}{n} \sum_{j=1}^{n} \left(\eta_j - \eta_m \right)^4 \tag{12.2}$$

where η_m and η_{rms} are the mean and root mean square values of the water surface elevation, respectively, and n is the number of recorded discrete data. Whereas the skewness describes the vertical asymmetry of the wave profile, the kurtosis indicates the extreme wave occurrence in the time series. The skewness and kurtosis correspond to 0 and 3 for the Gaussian linear waves.

Transient behavior of high-order nonlinear wave statistics in intermediate water

Here, we discussed the behavior of the high-order nonlinear wave statistics related to the extreme wave occurrence in intermediate water. The characteristics of the extreme wave occurrence in shallow water may differ from those of the extreme wave occurrence in deep water owing to the bottom effect. Fig. 12.2 shows the spatial evolution of the skewness and kurtosis of the water surface elevation in intermediate water. The horizontal axis is the distance from

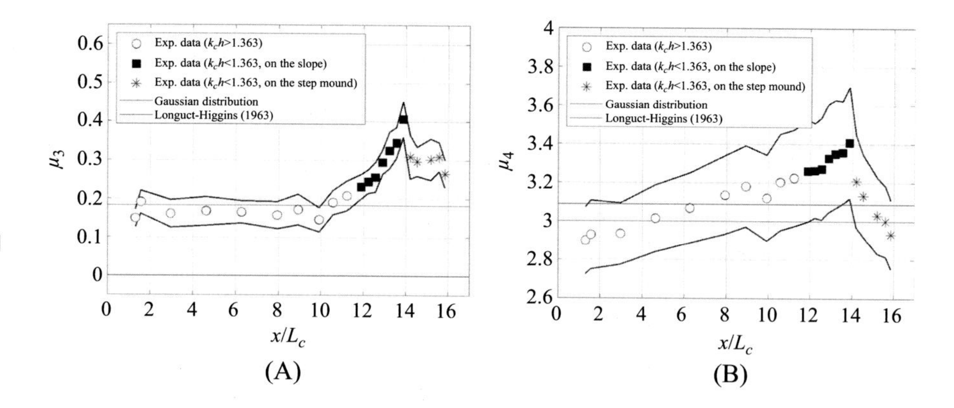

Figure 12.2 Spatial evolution of the skewness and kurtosis in the intermediate water (different marks: experimental data in different k_ch and bathymetry, dashed line: Gaussian distribution, solid line: second-order theory by Longuet-Higgins (1963), dotted lines: standard deviation). (A) Skewness and (B) kurtosis.

the wave maker x divided by the carrier wave wavelength $L_c = 2\pi/k_c$, and the leading edge of the slope started at $x/L_c = 8.0$. The vertical axes are the skewness and kurtosis derived from an ensemble average of 40 wave trains at each measurement point. The opened circle, filled square, and star marks indicate the experimental data for the same wave train at the location of $k_ch > 1.363$, $k_ch < 1.363$ on the slope and the step mound, respectively. The dotted lines are the standard deviation of 40 wave trains at each measurement point. We should note that the effect of breaking waves is not included for this condition at all the measurement points.

Note that $k_ch = 1.363$ means the critical condition of the Benjamin–Feir instability by Benjamin (1967). Yuen and Lake (1982) and Janssen and Onorato (2007) reported that the energy transfer and modulation of the quasi-monochromatic wave by the deepwater third-order nonlinearity occurred between four-wave components are diminished in the smaller water depth with $k_ch < 1.363$. The dashed line is the Gaussian distribution in which the skewness and kurtosis assume the value of 0 and 3, respectively. It is well known that the bound waves do not satisfy the linear dispersion relation and are caused by the second-order nonlinearity involving triads of waves (Longuet-Higgins, 1963). Longuet-Higgins (1963) developed the statistical property of second-order nonlinear random waves under the narrow-banded approximation (denoted the second-order nonlinear theory hereafter). In Fig. 12.2, the solid line corresponds to the expected values from the second-order nonlinear theory.

Based on Fig. 12.2, the distribution of the skewness deviates from the Gaussian distribution. This deviation is mainly dominated by the bound modes, even though the dynamics of waves can weakly contribute (Mori et al., 2007).

Therefore the skewness given by the second-order nonlinear theory is in good agreement with the experimental results for $k_c h > 1.363$. However, after passing the water depth with $k_c h = 1.363$ on the slope, the skewness rapidly increases up to 0.4 under the shallow water effect of the second-order nonlinear interactions. Furthermore, as waves propagate toward the step mound along the wave tank, the skewness drops to 0.3 and then becomes nearly constant on the step mound. Unlike skewness, kurtosis is more influenced by the nonlinear dynamics of free waves under the effects of third-order nonlinearity (Janssen, 2003; Mori and Janssen, 2006). From this point of view, the experimental data for $k_c h > 1.363$ shows that the kurtosis gradually increases as waves propagate along the wave tank and does not agree with both the second-order nonlinear theory and the Gaussian distribution. Even on the slope with $k_c h < 1.363$, the kurtosis smoothly increases up to 3.4. The aftereffect of the deepwater third-order nonlinearity to shallow water may remarkably influence the distributions of the random wave height and the maximum wave height in intermediate water. The evolution mechanism of the kurtosis from deep to shallow water and the effect on the wave height distribution are discussed in detail in Kashima and Mori (2019).

Application of standard Boussinesq equation to extreme wave modeling

Here, we investigate the application of the standard Boussinesq equation, which has been frequently and widely used to estimate wave transformation in shallow water (Hirayama, 2002), to high-order nonlinearity related to extreme waves. The standard Boussinesq equation shows high-level performance in the design of coast and harbor structures in Japan (Hirayama, 2013). The numerical simulations originally developed by Hirayama (2002) were performed to compare with the model experiments. The simulation is based on the standard Boussinesq equations with improved dispersion characteristics as reported by Madsen and Sørensen (1992). The governing equation is discretized by the alternating-direction implicit (ADI) method with the staggered grid, and the second-order central difference method and the Euler explicit method are applied to the spatial and temporal derivative terms, respectively.

The numerical domain was set up as shown in Fig. 12.3. z in this figure means the depth from the still water level. The wave signal was generated at the leftward boundary and propagated in the right-side direction in the numerical domain. The sponge layer was installed at the leftward boundary corresponding to the wave maker, and the permeable layer was installed at the end of the wave tank corresponding to the wave absorber. The detailed numerical conditions were the same as the experimental conditions. The computational spatial and temporal resolutions were set up as $dx = 0.05$ m and $dt = 0.001$ second to get high accuracy results of the numerical simulation, respectively.

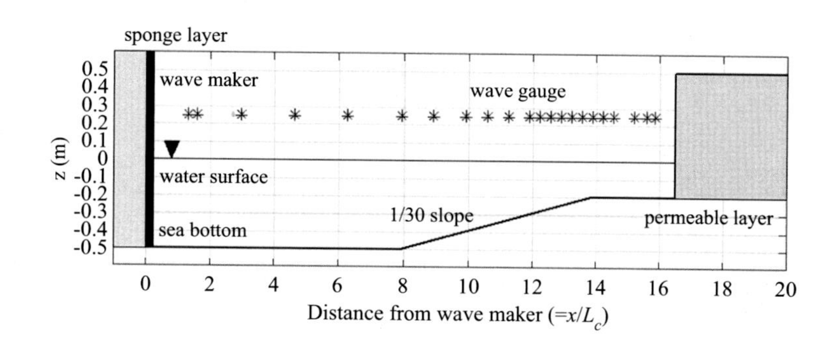

Figure 12.3 Sketch of model setup adopted in numerical simulation with the standard Boussinesq model developed by Hirayama (2002).

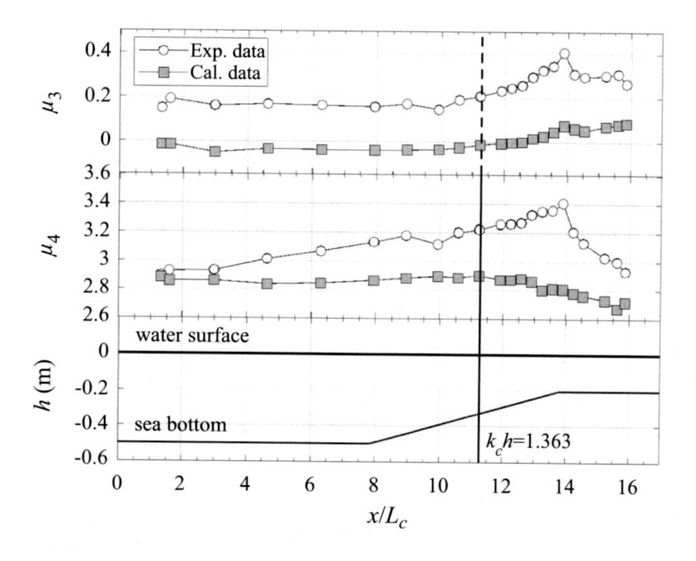

Figure 12.4 Comparison of the spatial evolution of the skewness and kurtosis in the intermediate water between model experiment and numerical simulations (circle: experimental data, square: calculated data).

Fig. 12.4 shows the comparison of the spatial evolution of the skewness and kurtosis in the intermediate water between the model experiment and numerical simulations. The circle and square indicate the experimental and calculated data, respectively. The spatial development of the skewness caused by the second-order nonlinearity involving triads of waves as shown in the experimental results for $k_c h > 1.363$ cannot be seen in the numerical simulation. The nonlinear interactions given by the standard Boussinesq equation are expressed by the balance of the nonlinear term of $O(\varepsilon)$ and the dispersion term of $O(\mu^2)$ where μ is characteristic relative water depth. Although the calculated skewness is slightly increased under the effect of the second-order nonlinear interactions with wave shoaling when the water depth becomes shallower on the slope and step mound, they remain less than one-fifth of the experimental data. On the other hand, the simulated kurtosis cannot appropriately show the spatial developments as shown in the experimental data. The kurtosis evolution caused by the four-wave interactions at the order of $O(\varepsilon^3)$ as shown in the experimental results, but such high-order nonlinear interactions of more than $O(\varepsilon,\mu^2)$ are not considered in the standard Boussinesq equation. Thus this result suggests that the standard Boussinesq equation may not appropriately evaluate the after-effect of the deepwater third-order nonlinearity to shallow water from the extreme wave modeling point of view.

High-order nonlinear effect on wave pressure acting on breakwater

Here, we demonstrate how offshore-generated wave groups by the four-wave interactions have an impact on the wave pressure acting on breakwater as the shallow water engineering application. There are several assumptions on the structural condition of a breakwater and the wave pressures acting on it. First, the Japanese standard composite breakwater model composed of a mound foundation and a vertical wall (e.g., caisson) is assumed to install at the same points as the wave gauges point on the step bottom in Fig. 12.1. The model experiments were performed without any breakwater models in the wave tank in the following general way because it is necessary to get the progressive wave properties to estimate the wave pressures using Goda's formula (1973). Second, the existence of impulsive wave pressures and wave breaking around the breakwater is not considered to avoid additional effects for simplicity.

The wave pressures acting in front of the vertical wall of the breakwater have a trapezoid-shape distribution, according to Goda (1973) as shown in Fig. 12.5. The wave pressure on the still water level p_1, denoting the wave pressure simply hereafter, is regarded as the representative value of the wave pressure distribution along the vertical wall and is calculated by using the wave pressure formula proposed by Goda (1973) as the following equations:

$$p_1 = \frac{1}{2}(1+\cos\beta)\left(\alpha_1\lambda_1 + \alpha_2\lambda_2\cos^2\beta\right)\rho_w g H_D \tag{12.3}$$

$$\alpha_1 = 0.6 + \frac{1}{2}\left(\frac{2k_t h_t}{\sinh(2k_t h_t)}\right)^2 \tag{12.4}$$

$$\alpha_2 = \min\left\{\frac{h_b - d}{3h_b}\left(\frac{H_D}{d}\right)^2, \frac{2d}{H_D}\right\} \tag{12.5}$$

where H_D is defined as the maximum wave height of the progressive wave train; ρ_w and g are the density of water and gravity acceleration, respectively; β is the angle between the direction of wave propagation and a line normal to the breakwater and here $\beta = 0$; h_t denotes the water depth in front of the breakwater; d denotes the depth above the mound foundation; k_t denotes the wave number

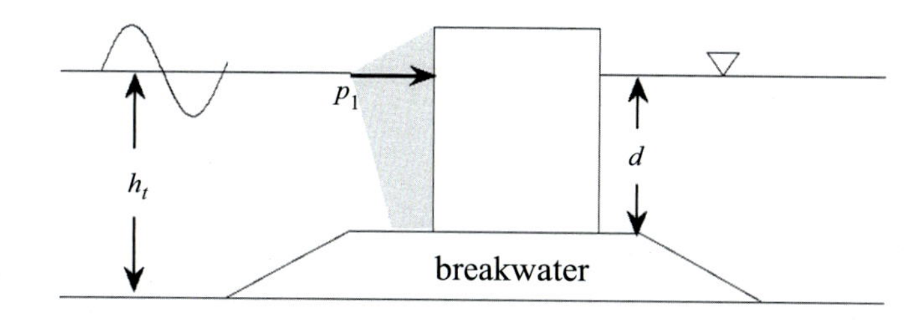

Figure 12.5 Structure condition of a breakwater model (p_1: wave pressure on still water level, d: depth above the mound foundation, h_t: water depth in front of the breakwater).

at the water depth h_t; h_b denotes the water depth at the location at a distance five times the significant wave height seaward of the breakwater; and λ_1 and λ_2 are the coefficients related to the structure, and $\lambda_1 = \lambda_2 = 1.0$ for the composite breakwater. These formulae are derived under the assumption that the wave height distribution is approximated by the Rayleigh theory for free waves when the wave breaking does not occur around the breakwater (Goda, 1973).

The wave pressures acting on the breakwater were estimated by using three different calculation methodologies of the maximum wave height. The first and second methods are (A) direct use of H_{max} and (B) use of 1.8 times $H_{1/3}$ obtained from the statistical analysis with $N = 250$ for the experimental measurements, respectively. The second method gives variations of the wave pressure based on the changes of $H_{1/3}$ (i.e., wave energy). The last one, method (C), is the maximum wave height given by the Japanese design method for a breakwater taking into account the nonlinear wave shoaling effect. In this method, first $H_{1/3}$ is estimated by the product of the offshore wave height, which corresponds to the initial wave height for the experiments, and the nonlinear wave shoaling coefficient based on Shuto (1974). Then, H_{max} can be given as 1.8 times the above $H_{1/3}$ according to the Rayleigh theory. Method (C) gives a single value for wave pressure distribution without statistical and dynamic fluctuations. Hereafter, we call the wave pressure given by the last methods as "designed wave pressure $(p_1)_d$" according to Eq. (12.3). In order to investigate the effect of changing the occurrence frequency of the maximum wave height on the wave pressure p_1, Fig. 12.6 shows the probability density function (PDF) of the wave pressure using the three maximum wave heights on the slope and step mound. Each wave pressure is normalized by the designed wave pressure $(p_1)_d$. The dotted line with the filled circle and the dotted-dashed line are the distributions of the wave pressure using the H_{max} (A) and 1.8 times of $H_{1/3}$ (B), respectively. The vertical solid line corresponds to the designed wave pressure (C). The wave pressure using H_{max} (A) for the experimental data is distributed at the lower-pressure side than the designed wave pressure on the slope as shown in Fig. 12.6A. In addition, there are about $\pm 20\%$ fluctuations by method (A) in the developed nonlinear situation,

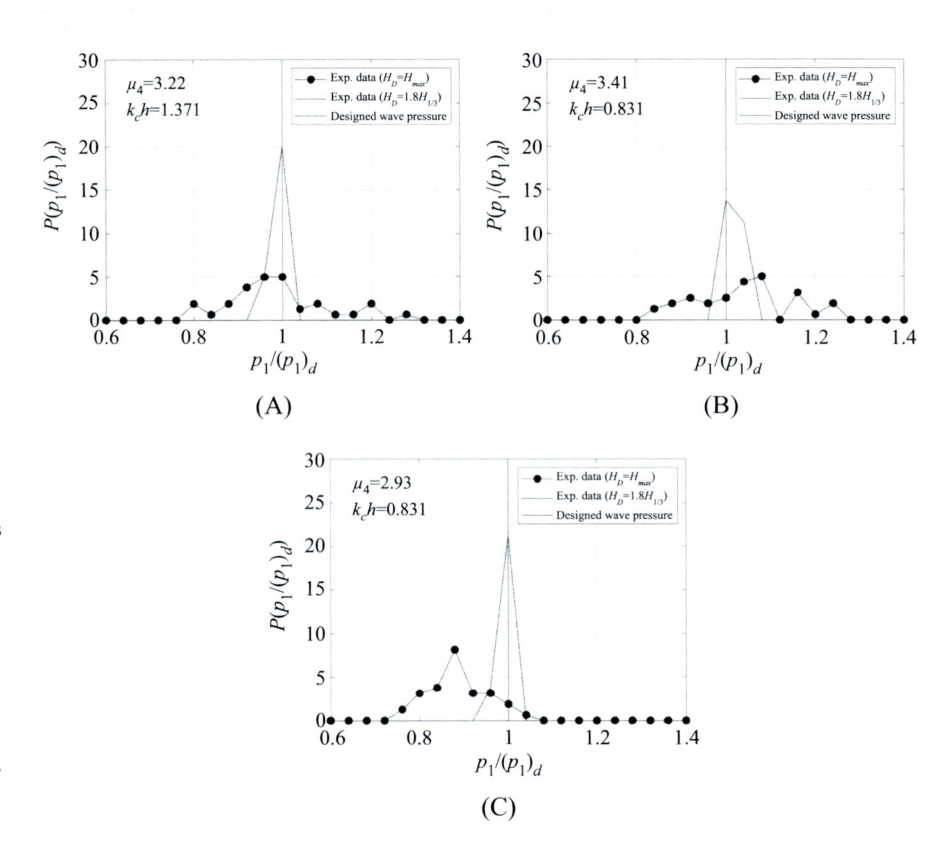

Figure 12.6 PDF of the wave pressure acting on the breakwater (dotted line with filled circle: experimental data based on H_{max} [method A], dotted-dashed line: experimental data based on $H_{1/3}$ [method B], vertical solid line: standard design value based on offshore $H_{1/3}$ [method C]). (A) *w*10 (on the slope), (B) *w*17 (at the edge of slope), and (C) *w*22 (on the step mound).

although method (B) gives only $\pm 5\%$ fluctuations due to the stability of $H_{1/3}$ in comparison with $H_{\max}$. After propagating on the slope with $k_c h < 1.363$, the wave pressure is more widely distributed exceeding the designed wave pressure. This behavior can be seen even on the edge of the step mound where $k_c h$ is much less than 1.363 (Fig. 12.6B). Finally, the wave pressure is distributed at the lower-pressure side than the designed wave pressure, and its peak shifts to smaller wave pressures on the step mound (Fig. 12.6C). Unlike the distributions using the $H_{\max}$, the variation of the wave pressure using 1.8 times of $H_{1/3}$ is very small, and its peak is nearly constant with the designed wave pressure regardless of the water depth changing. Therefore, as the high-order nonlinearity of the waves becomes stronger, the wave pressure in shallow water with $k_c h < 1.363$ tends to exceed the designed wave pressure depending on the after-effect of the high-order nonlinearity for individual wave trains in deeper water.

Finally, Fig. 12.7 shows the dependence of the wave pressure as a function of the kurtosis without considering relative water depth. The dark black and light gray boxes show the experimental results $(p_1)_{H\max}$ from method (A) and $(p_1)_{1.8H1/3}$ from method (B), respectively. Each wave pressure is normalized by

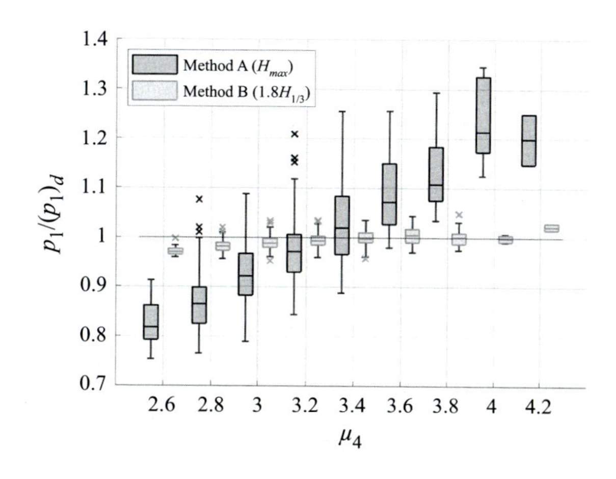

Figure 12.7 Variation of the wave pressures acting on the breakwater as a function of the kurtosis (black: maximum pressure based on H_{max} normalized by standard design pressure [method A], gray: maximum pressure based on $H_{1/3}$ normalized by standard design pressure [method B]).

the designed wave pressure (C). The box indicates the interquartile range (IQR), which is a measure of statistical dispersion. The ends of the box correspond to the first and third quartiles, respectively. The line in the box is the median, and the lines extending vertically from the box, that is, whiskers, indicate variability outside the first and third quartiles. Each end of the whiskers is the maximum and minimum values of the data within 1.5 IQR of the first and third quartiles, respectively. The x-mark indicates the data without 1.5 IQR of the first and third quartiles. The wave pressure using 1.8 times of $H_{1/3}$ (B) follows the behavior of the designed wave pressure and is nearly constant regardless of the kurtosis. This is because kurtosis has no direct effect on the significant wave height. On the other hand, the variation of the wave pressure using H_{max} (A) is larger than that of the wave pressure using 1.8 times of $H_{1/3}$ (B). Moreover, the wave pressure using H_{max} (A) is increased with the kurtosis changing. As a result, it exceeds the designed value around $\mu_4 = 3.4$ on average, while the Rayleigh theory which is used for the structure design assuming $\mu_4 = 3.0$ slightly overestimates it. Note that this result indicates both under- and overestimation of maximum wave pressure acting on the breakwater for the assumed condition here. Therefore it is possible to evaluate the uncertainty of the wave pressure acting on the breakwater under the same wave energy appropriately by using the kurtosis changing related to the occurrence frequency on maximum wave height.

Conclusion

A series of model experiments in a wave tank installed with a step bottom model was conducted to investigate the shallow water effect on the extreme wave occurrence in intermediate water and the wave pressures acting on the breakwater under the extreme wave conditions for the unidirectional random waves. As a result of our investigation, the threshold value for instability $k_c h = 1.363$ plays a significant role in understanding the extreme wave occurrence from deeper to shallow water. While the dependence of the kurtosis on the extreme wave occurrence is weakened due to the reduction of the nonlinear energy transfer for sufficiently small water depth, $k_c h < 1.363$ (Benjamin, 1967), the influences of the deepwater third-order nonlinearity in greater water depth $k_c h > 1.363$ remain even in shallow water depending on the bottom bathymetry. Furthermore, from the point of view of the design for the breakwater, it is found that there is a clear dependency of the kurtosis on the wave pressures acting on the breakwater and their variation.

References

Benjamin, T.B., 1967. Instability of periodic wavetrains in nonlinear dispersive system. Proceedings of the Royal Society A A299, 59–75.

Goda, Y., 1970. Numerical experiments on wave statistics with spectral simulation. Report of the Port and Harbour Research Institute 9 (3), 3–57.

Goda, Y., 1973. A new method of wave pressure calculation for the design of composite breakwaters. Report of the Port and Harbour Research Institute 12 (3), 31–69 (In Japanese).

Goda, Y., 2000. Random Seas and Design of Maritime Structures. World Scientific.

Hasselmann, K., et al., 1973. Measurements of wind-wave growth and swell decay during the joint north sea wave project (JONSWAP). Deutschen Hydrographischen Zeitschrift Reihe A 8, 12.

Hirayama, K., 2002. Utilization of numerical simulation on nonlinear irregular wave for port and harbor design. Technical Note of the Port and Airport Research Institute 1036, 162 (in Japanese).

Hirayama, K., 2013. Harbor tranquility analysis method for using Boussinesq-type nonlinear wave transformation model. In Proceedings of 23rd International Offshore and Polar Engineering Conference, Rhodes, Greece.

Itoh, Y., Fujishima, M., Kitatani, T., 1966. On the stability of breakwater. Report of the Port and Harbour Research Institute 5 (14), 134 (In Japanese).

Janssen, P.A.E.M., 2003. Nonlinear four-wave interactions and freak waves. Journal of Physical. Oceanography 33 (4), 863–884.

Janssen, P.A.E.M., Onorato, M., 2007. The intermediate water depth limit of the Zakharov equation and consequences for wave prediction. Journal of Physical. Oceanography 37 (10), 2389–2400.

Kashima, H., Hirayama, K., Mori, N., 2013. Numerical study of aftereffects of offshore generated freak waves shoaling to coast. Proceedings of the 7th Coastal Dynamics 2013, 947–956.

Kashima, H., Hirayama, K., Mori, N., 2014. Estimation of freak wave occurrence from deep to shallow water regions. Proceedings of the 34th International Conference of Coastal Engineering 1 (34), 36. Available from: https://doi.org/10.9753/icce.v34.waves.36.

Kashima, H., Mori, N., 2019. Aftereffect of high-order nonlinearity on extreme wave occurrence from deep to intermediate water. Coastal Engineering 153, 103559. Elsevier.

Longuet-Higgins, M., 1963. The effect on nonlinearities on statistical distributions in the theory of sea waves. Journal of Fluid Mechanics 17, 459–480.

Madsen, P.A., Sørensen, O.R., 1992. A new form of the Boussinesq equations with improved linear dispersion characteristics. Part 2. A slowly-varying bathymetry. Coastal Engineering 18, 183–204.

Mendez, F.J., Losada, I.J., Medina, R., 2004. Transformation model of wave height distribution on planar beaches. Coastal Engineering 50 (3), 97–115. Available from: https://doi.org/10.1016/j.coastaleng.2003.09.005.

Mori, N., Janssen, P.A.E.M., 2006. On kurtosis and occurrence probability of freak waves. Journal of Physical. Oceanography 36 (7), 1471–1483.

Mori, N., Onorato, M., Janssen, P.A.E.M., Osborne, A.R., Serio, M., 2007. On the extreme statistics of long-crested deep water waves: theory and experiments. Journal of Geophysical Research 112, C09011. Available from: https://doi.org/10.1029/2006JC004024.

Shuto, N., 1974. Nonlinear long waves in a channel of variable section. Coastal Engineering in Japan 17, 1–12.

Trulsen, K., Zeng, H., Gramstad, O., 2012. Laboratory evidence of freak waves provoked by non-uniform bathymetry. Physics of Fluids 24, 097101.

Whitham, G.B., 1974. Linear and Nonlinear Waves. Wiley.

Yuen, H., Lake, B.M., 1982. Nonlinear dynamics of deep-water gravity waves. Advances in Applied Mechanics 22, 67–327.

Zeng, H., Trulsen, K., 2012. Evolution of skewness and kurtosis of weakly nonlinear unidirectional waves over a sloping bottom. Natural Hazards and Earth System Sciences 12, 631–638.

Zhang, J., Benoit, M., Kimmoun, O., Chabchoub, A., Hsu, H.C., 2019. Statistics of extreme waves in coastal waters: large scale experiments and advanced numerical simulations. Fluids 4 (99), 1–24.

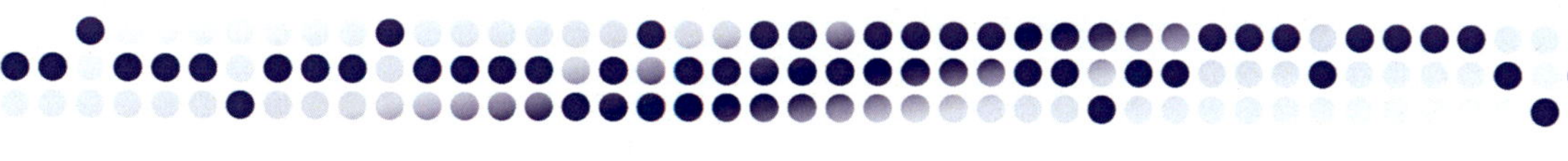

Application 4: controlled freak wave generation and recreation in hydrodynamic facilities

Amin Chabchoub[1,2]
[1]Disaster Prevention Research Institute, Kyoto University, Japan [2]School of Civil Engineering, The University of Sydney, Sydney, NSW, Australia

The need to study extreme waves in a controlled laboratory environment

Ocean rogue waves have been reported in all parts of the oceans and confirmed by several measurements (Chapters 1–4). That said, encountering an extreme wave in the ocean is still rare, and most of the in-situ rogue wave measurements have been captured coincidentally (Kharif et al., 2009). These time series have been recorded by downward-looking lasers, installed on offshore platforms, or accelerometers/inclinometers of a buoy, and this makes quantification of the wave kinematics or detection of the focusing origin not possible. Spatiotemporal measurements using stereo imaging techniques, as presented in Chapter 4, are promising since more accurate physical features of rogue waves can be captured, identified, and quantified (Malila et al., 2023).

On the other hand, the design of maritime vehicles and structures requires prolonged testing to meet stability and safety standards (Chapters 10–12). These cannot be easily conducted in an open ocean environment, and water wave facilities provide excellent settings to study wave–structure interaction problems and key dynamic properties of freak waves.

State-of-the-art water wave facilities include either a single or several computer-controlled wave makers, which can generate predefined time series or reference narrowband/broadband sea states with random phases, such as JONSWAP-Pierson–Moskowitz-type wave spectra (Holthuijsen, 2010). To accurately reproduce waves' orbital motion, flap- or hinged-type wave makers are suitable to

Science and Engineering of Freak Waves.
DOI: https://doi.org/10.1016/B978-0-323-91736-0.00009-2

generate deep water waves, whereas piston-type paddles are used in finite depth and shallow water environments. Worthy of mention is that modern wave makers account for the second-order wave generation steering motion. If wave reflection is not desired, a wave absorber is placed at the opposing end to attenuate the wave energy accordingly. Capacitance, resistive, or ultrasound wave gauges are commonly installed along the propagation direction of the waves to track their evolution while force transducers/wave load cells on marine model structures measure the respective wave forces/loads on predesigned mechanical structures. Since the wave gauge data acquisition is not defying, it is more common to study temporal variations of surface elevation along the spatial evolution direction(s). Then again, it is challenging to use stereo imaging due to light reflections in the laboratory while spatiotemporal measurements of wave dynamics using an array of markers or wave gauges remain time-consuming (Chabchoub et al., 2019; Steer et al., 2019). Particle image velocimetry or particle tracking velocimetry are customarily used to measure the flow velocity and quantify wave-breaking processes (Alberello et al., 2018; van den Bremer et al., 2019).

Such well-controlled equipment allows for the perfect repeatability of water wave experiments and comprehensive investigation of extreme waves in a "bathtub" in comparison to ocean scales. Fig. 13.1 shows two examples of state-of-the-art wave facilities: a unidirectional wave flume and a wave basin.

Excluding any external effects, i.e., bathymetry change, wind, or currents, rogue waves can be generated in the laboratory following three fundamental mechanisms, which have been already mentioned in Chapters 1 and 2. These are the interference, modulation instability (MI), and time-reversal principles. Fig. 13.2 illustrates these three mechanisms while associated experiments will be discussed below in the following sections.

There are landmark review papers discussing ocean freak waves and highlighting experimental investigations in wave tanks. We refer the reader to

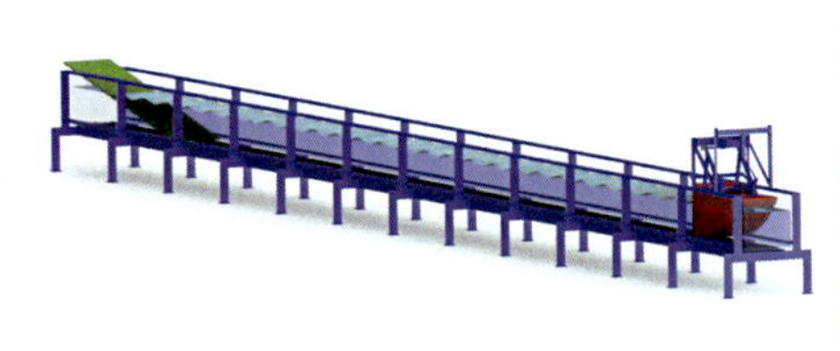

Figure 13.1 Left: Schematic view of a unidirectional water wave tank as installed at the University of Sydney with dimensions $30 \times 1 \times 1$ m^3, comprising a wave maker and a removable wave-absorbing beach. Right: Japan's National Maritime Research Institute (NMRI) Actual Sea Model Basin with the dimensions $80 \times 40 \times 4.5$ m^3. The segmented wave makers operate as wave generators or active absorbers. *Source: The picture was taken from the NMRI website.*

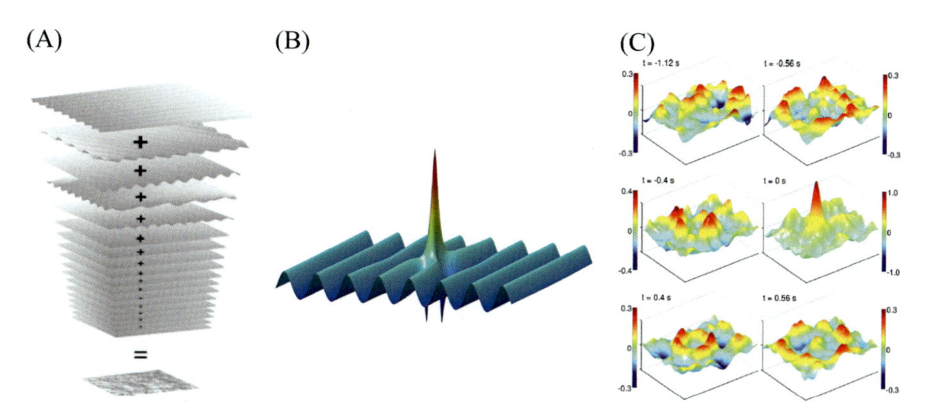

Figure 13.2 Illustration of: (A) linear interference principle. (B) Exemplified modulation instability focusing process arising on a periodic wave envelope. (C) Time-reversal focusing of water waves. *Source: Reprinted with permission from (A) Holthuijsen, L.H., 2010. Waves in Oceanic and Coastal Waters. Cambridge University Press; (C) Przadka, A., Feat, S., Petitjeans, P., Pagneux, V., Maurel, A., Fink, M., 2012. Time reversal of water waves. Physical Review Letters 109, 064501.*

Kharif and Pelinovsky (2003), Dysthe et al. (2008), Onorato et al. (2013), Adcock and Taylor (2014), Dudley et al. (2019), Sapsis (2021), and Ma et al. (2022b) for a comprehensive overview.

Wave interference

The superposition principle, also known as dispersive focusing, is probably the most intuitive explanation for rogue wave formations in the ocean. In a very simplistic way, ocean surface waves can be understood as a sum of different harmonic waves interfering with each other (Pierson, 1955; Holthuijsen, 2010). The temporal surface displacement with random phase φ, which can be generated by a wave maker at the fixed location x^*, is parametrized as

$$\eta(x^*, y, t) = \sum_{i=1}^{n}\sum_{j=1}^{m} a_{i,j}\cos\Big(\omega_i t - k_i x^* \cos\vartheta_j - k_i y \sin\vartheta_j + \varphi_{i,j}\Big). \tag{13.1}$$

That is, in addition to a one-dimensional model with wave frequency ω and random phase φ, we now include the directional spreading angle ϑ. When considering the linear dispersion relation in finite water depth $\omega^2 = gk \tanh kh$, it is clear that a wave with a smaller frequency value travels faster than its counterpart with a higher frequency. This remains evidently true in deep water with the dispersion relation becoming $\omega^2 = gk$. As such, rogue waves can be formed because of an unfortunate circumstance of phase matching and resulting amplitude overlap (Longuet-Higgins, 1974; Baldock et al., 1996).

However, when generating extreme waves in the laboratory, this simplified approach, which is based on linear wave theory, fails in appointing the exact focusing location in a water wave facility. The reason for this is trivial since waves are by nature nonlinear (Kharif et al., 2009; Osborne, 2010; Babanin, 2011). Considering, for instance, the power spectrum of a generated unidirectional and regular wave field, the latter perpetually manifests bound waves or higher Stokes wave contributions in the form of energy peaks at $2f$, $3f$, $4f$, etc. The value of wave steepness ak, which is also a scaling parameter, controls the magnitude of these respective contributions. Hence, when considering small amplitude waves in the modeling, it may be sufficient to adopt the linear theory when focusing waves based on the superposition principle. However, a second-order correction in the phase velocity is indispensable if attempting to superimpose waves in large wave tanks with significant fetch (Alberello et al., 2018; Fedele et al., 2016; Gemmrich and Cicon, 2022). We refer to the recent review paper by Ma et al. (2022b) for a summary, particularly on the extreme and breaking waves resulting from dispersive focusing in water wave facilities.

It is noteworthy to mention that the famed Draupner or New Year wave, which is discussed in Chapter 1, has been created on the basis of the superposition principle in a unidirectional tank (Clauss and Klein, 2011), and as represented in the top panel of Fig. 13.3. A more realistic recreation including breaking attributes in a directional basin and following the crossing sea hindcast has been reported by McAllister et al. (2019), see bottom panel of Fig. 13.3.

Indeed, it is nowadays well-known that the underlying meteorological conditions suggest that two crossing wave systems collided around the location of the Draupner platform. Therefore, it is reasonable to hindcast and conjecture that the New Year wave occurred in a cross sea state (Cavaleri et al., 2016).

Nonlinear focusing

This generation mechanism is based on the instability of Stokes waves subject to long-wave perturbations. In their pioneering work, Benjamin and Feir (1967) applied a linear stability analysis to sideband-perturbed second-order Stokes waves. They showed that for a given wave train with wave frequency ω and wave steepness ak, a small amplitude sideband disturbance around the peak frequency will grow exponentially, if triggered within the frequency range

$$0 < \Omega < \sqrt{2}ak\omega. \tag{13.2}$$

The resulting wave focusing and distortion have been observed and reported in the same year by Benjamin (1967).

The identical MI criterion was found independently after application of linear stability on the regular wave envelope within a framework of *simplified* weakly nonlinear framework, which is derived from the Zakharov equation assuming a narrowband process at play (Zakharov, 1968). The latter third-order approximative wave evolution equation was initially referred to as the cubic equation.

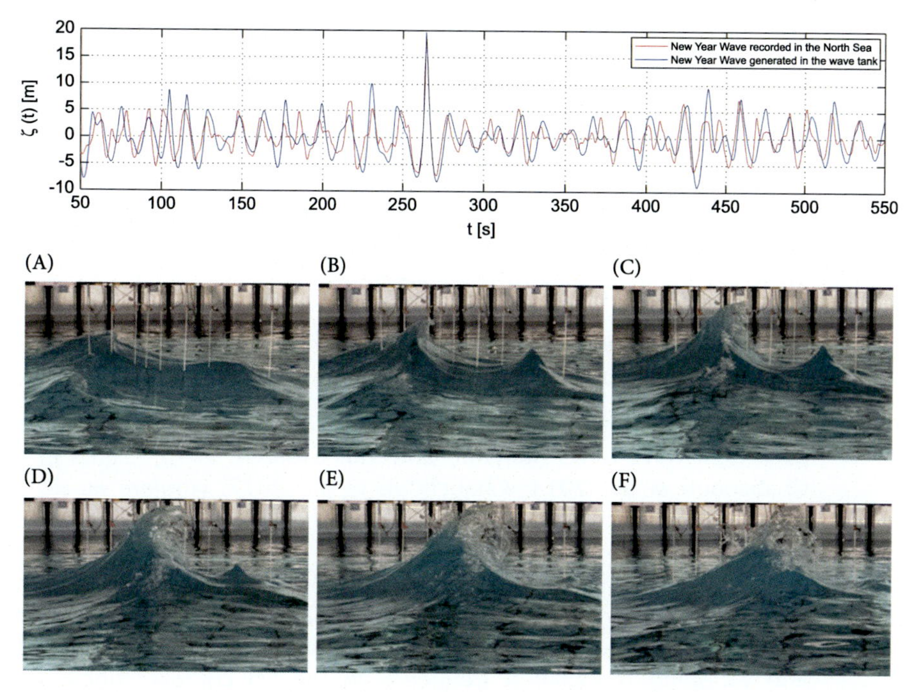

Figure 13.3 Draupner wave recreation following the dispersive focusing principle. Top panel: generation in a unidirectional tank. Bottom panels (A)–(F): reconstruction in a wave basin according to the meteorological hindcast. *Source: (Top panel) Reprinted with permission from Clauss, G.F., Klein, M., 2011. The new year wave in a seakeeping basin: Generation, propagation, kinematics and dynamics. Ocean Engineering 38, 1624–1639; (Bottom panels) From McAllister, M.L., Draycott, S., Adcock, T., Taylor, P., van den Bremer, T., 2019. Laboratory recreation of the Draupner wave and the role of breaking in crossing seas. Journal of Fluid Mechanics 860, 767–786.*

However, due to its similarity to the quantum Schrödinger equation, it has been referred to as the nonlinear Schrödinger equation (NLSE). In deep water, it reads

$$i\left(\psi_x + \frac{2k}{\omega}\psi_t\right) - \frac{k}{\omega^2}\psi_{tt} - k^3|\psi|^2\psi = 0. \tag{13.3}$$

The advantage of analyzing the MI within the NLSE framework is twofold. The description of modulationally unstable wave fields is more physical since the framework predicts a saturation of the wave focusing process, which is then followed by a subsequent decay of the wave amplitude (Lake et al., 1977; Tulin and Waseda, 1999). Furthermore, the NLSE possesses a wide range of exact envelope expressions, which describe different MI scenarios (Akhmediev and Ankiewicz, 1997; Dysthe and Trulsen, 1999). This makes the manipulation and control of a variety of unstable wave groups in a hydrodynamic facility much easier to handle, compared to the classic MI excitation by a three-wave system (one peak frequency and a pair of side bands). Consequently, the rogue surface wave profiles can be accurately generated whereas the location of expected maximal wave focusing can be translated in time or space as desired. The boundary conditions, as imposed by the wave envelope ψ and applied to the wave maker, are defined by the expression of surface elevation.

$$\eta(x^*, t) = \mathrm{Re}\big(\psi(x^*, t)\exp[i(kx^* - \omega t)]\big). \tag{13.4}$$

Note that the linear surface parameterization is sufficient to trigger such rogue wave envelope dynamics. The wave maker's transfer function should be accurately calibrated to allow the generation of exact wave amplitudes, as determined by the complex wave envelope $\psi(x, t)$, which contains not only the information of wave amplitude modulation but also of the crucial localized phase-shift perturbation (Osborne, 2010; He et al., 2022a).

Several studies dealt with the controlled generation of periodic unstable wave packets in wave tanks (Karjanto and Van Groesen, 2010; Clauss et al., 2011; Chabchoub et al., 2014; Chabchoub et al., 2017; Houtani et al., 2018). In effect, there is a consensus that the fundamental breather, which qualitatively resembles the dynamics of ocean rogue waves, is the Peregrine breather (Peregrine, 1983; Shrira and Geogjaev, 2010; Kibler et al., 2010; Chabchoub et al., 2011). There are some remarkable features of this pulsating wave envelope. As a matter of fact, it is a particular case of an infinite modulation period and vanishing modulation frequency in the MI context. Since the doubly-localized Peregrine breather amplifies the amplitude of the wave field by a factor of three, it also violates the narrowband assumption adopted in the derivation of the NLSE. Remarkably, a Peregrine-type perturbation qualitatively evolves as predicted by the NLSE theory, despite all physical limitations of the modeling and inevitable experimental dissipative effects (Fig. 13.4).

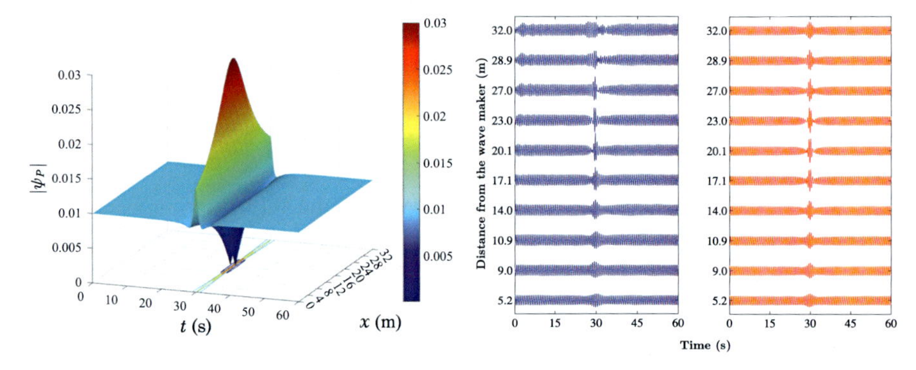

Figure 13.4 Evolution of a Peregrine breather for $a = 0.01$ m and $ak = 0.09$ as traveling with the value of group velocity c_g. Left panel: theoretical prediction of the wave envelope dynamics. Middle panel (blue lines): surface elevation evolution as measured in the University of Tokyo's wave tank. Right panel (red lines): NLSE surface elevation prediction.

An improved agreement with nonlinear wave evolution forecast can be obtained by taking the perturbation analysis one step further, i.e., to the fourth-order of approximation. This framework is well-known as the modified NLSE or simply the Dysthe equation (Dysthe, 1979). Associated validation studies have been conducted, for instance, by Shemer and Alperovich (2013) and Slunyaev et al. (2013).

Being a deterministic and controllable rogue wave model, the Peregrine breather is now used as a reference model to investigate unstable rogue wave group dynamics and to assess their impact on marine structures (Klein et al., 2016; Deng et al., 2016; Liao et al., 2018).

In addition, wave breaking kinematics of unstable wave groups substantially differ from extremes resulting from the superposition principle. The gradual increase of wave amplitude within the wave group and the recurrent breaking behavior (Chapter 2) of breather-type wave packets in particular, ensures the long survivability of an extreme wave event, which involves a spectral downshifting, distinctive asymmetry in the flow velocity, and substantial dissipation in the air (Tulin 1996; Iafrati et al., 2013; Shemer and Liberzon, 2014; Peric et al., 2015; Eeltink et al., 2017; Alberello et al., 2018).

The latest experimental progress underlines the fact that MI has much richer features beyond simple excitation within a narrowband wave field and as implied by the linear stability analysis. Examples are doubly periodic large-amplitude breathers that can be excited by modulation frequencies outside the classical frequency range of Eq. (13.2) (Eeltink et al., 2020; Vanderhaegen et al., 2021). Moreover, the evolution of modulationally unstable coherent wave groups appears to be robust to strong wave field perturbations and can be either embedded in a JONSWAP-type sea state (Chabchoub et al., 2017) or spontaneously emerging from either noisy or irregular wave fields (Chabchoub et al., 2017; Michel et al., 2020).

In addition, laboratory experiments confirmed the robustness of coherent breather wave groups to strong wind perturbations (Chabchoub et al., 2013). Breathers can also have a directional coherence (Chabchoub et al., 2019). Indeed, similar oblique ocean wave groups have striking similarities (Waseda et al., 2021). Recently, it has been demonstrated that breathers can also evolve and propagate without any disturbance in standing wave systems, which is a particular case of a cross wave system (He et al., 2022b). Before then, scenarios in which the MI is involved in rogue wave formation in crossing seas have been ruled out. Future studies are required to systematically investigate the stability of unstable and coherent wave group propagation when colliding with other wave systems by accounting for diverging cross angles as well as spectral bandwidths.

The investigation of the role of third-order nonlinear effects or quasi-resonant interactions in the formation of rogue waves in irregular sea states was tackled in several experimental campaigns in unidirectional wave tank since 2006 (Onorato et al., 2006; Dematteis et al., 2019; Suret et al., 2020). In these latter experimental studies, the critical role of second- and third-order nonlinear effects, and breathers in rogue wave statistics has been investigated. Detection of non-Gaussianity, and departure of kurtosis from the typical Gaussian distribution value of three, can be easily achieved in such conditions by either narrowing the wave spectrum, increasing the characteristic steepness, or both. This is well quantified by the Benjamin–Feir index (Chapters 2 and 5).

When including the directional wave spreading in the modeling, the work by Waseda et al. (2009) confirmed the weakening of the nonlinear third-order effects and the reduction in the frequency of rogue wave events when the spreading angle is larger than 20 degrees. This leads to the dominance of second-order effects in the generation of freak waves in such setups. Similar results have been reported by Onorato et al. (2009), Latheef and Swan (2013), Karmpadakis et al. (2019), and Michel et al. (2022). All these experimental works agree with recent space-time field measurements in the North Sea reported by Malila et al. (2023). These long-term measurements also confirmed that wind-dominated sea states, characterized by a narrow directional spreading, show signatures of nonlinear wave group dynamics.

Time-reversal recreation

The time-reversed refocusing of pulses is a renowned mechanism widely adopted in acoustics as well as imaging and is a highly efficient principle to control wave focusing (Fink, 1997). It can be, for instance, observed in our daily life by watching a filter coffee machine's carafe. Each falling coffee drop creates concentric waves, which are then reversed after hitting the carafe's wall, or technically mirror within the time-reversal context. Once these reversed waves reach the center of the carafe, it is fascinating to observe the drop reforming on the surface.

This principle is at the same time "simple" as well as elegant and has far-reaching physical implications. More importantly, it applies to both, linear and nonlinear waves, as long as the dissipation processes involved are either negligible or minor and quasi-linear. It is also worth mentioning that a rapid and sudden change in the gravity time-reverses water waves (Bacot et al., 2016).

Reported experiments on time-reversal refocusing for water waves have been proposed for the first time for linear waves (Przadka et al., 2012) and later for breathers, which are strongly nonlinear waves (Chabchoub and Fink, 2014). In this latter work, it is shown that the reciprocity of the NLSE Eq. (13.3) allows reinjecting the time-reversed wave gauge signal at the wave maker location, instead of the mirror location. Consequently, it is expected to have a refocusing at the mirror position, rather than at the wave generator position.

This fundamental property has been proven to be very efficient in reconstructing real-world ocean rogue wave profiles at smaller amplitude ratios, since by applying the time-reversal procedure, the initially generated extreme wave can be refocused at any desirable distance from the wave maker following the application of the time-reversal procedure (Ducrozet et al., 2016). More specifically, the Draupner and Yura waves have been successfully reconstructed in a large unidirectional facility based on this principle (Ducrozet et al., 2020). Fig. 13.5 shows an example of the reconstructed Draupner wave for four different adopted amplitude ratios.

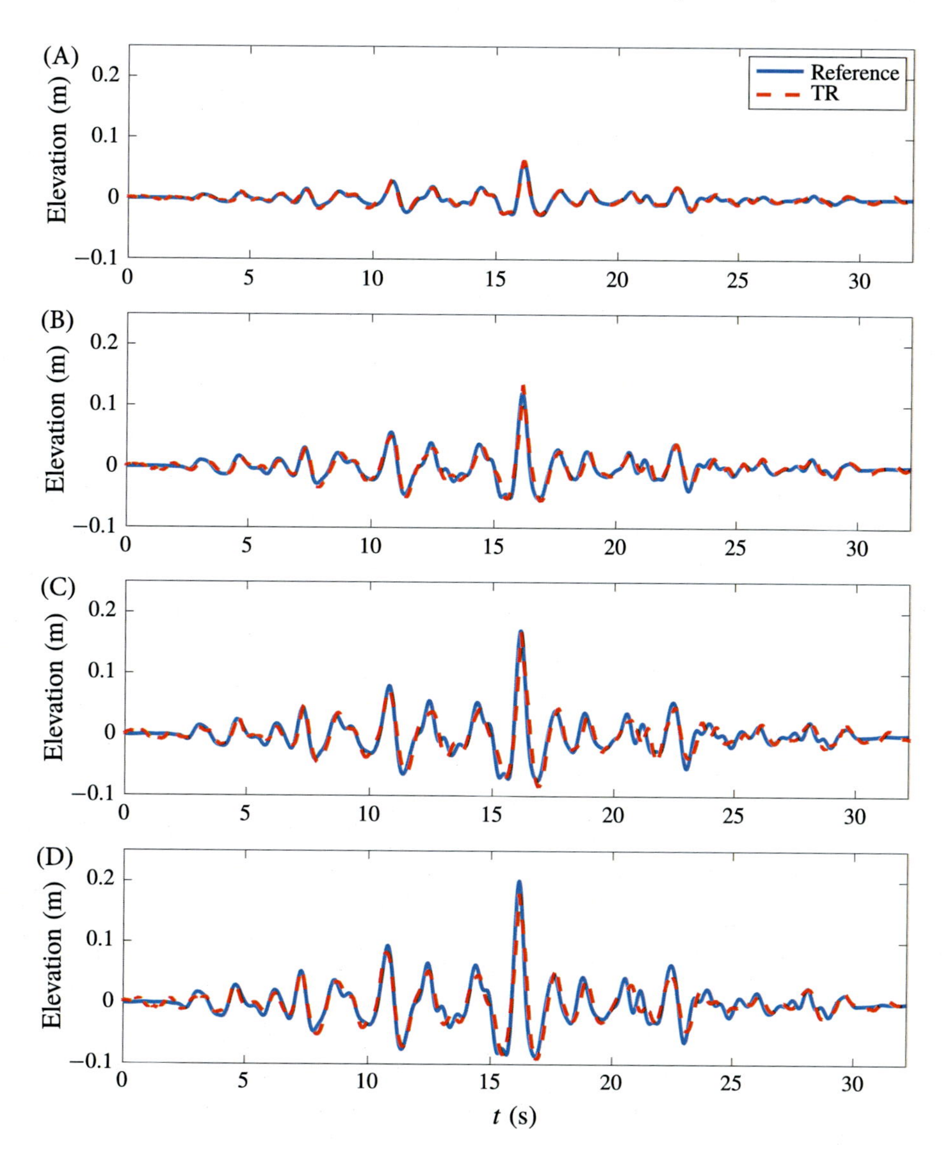

Figure 13.5 Comparison of the measured (solid lines) and reconstructed (dashed lines) Draupner wave using the time-reversal principle adopting four amplitude ratios and as measured 15 m from the wave maker. *Source: Reprinted with permission from Ducrozet, G., Bonnefoy, F., Mori, N., Fink, M., Chabchoub, A., 2020. Experimental reconstruction of extreme sea waves by time reversal principle. Journal of Fluid Mechanics 884, A20.*

Additional in-situ freak waves have been recreated with corrections in wave generation as well as amplitude and phase (Ma et al., 2022a). Irregular and directional waves have also been reconstructed in wave basins using this technique (de Mello et al., 2016; Draycott et al., 2022).

It is obvious that wave breaking and substantial dissipation are natural limitations for the application of the time-reversal principle. But as long as these remain marginal, the time-reversal wave focusing has been shown to be a very efficient and effective technique for the controlled recreation of rogue waves since it can be applied to narrowband and broadband as well as linear and nonlinear wave processes.

Perspectives

Considering the recent impressive progress in experimental nonlinear extreme wave hydrodynamics as well as the new physical insights gained from these, it is anticipated that a strong continuation of this trend will reveal further unknown and decisive features of rogue wave dynamics, particularly in multidirectional sea states.

It is evident that laboratory investigations are essential to confirm novel proof of concepts and determine the limitations of the respective physical modeling. Moreover, a comprehensive understanding of the wave interactions with the bottom topography (Chapter 7), wind (Kharif et al., 2008; Toffoli et al., 2017; Geva and Shemer, 2022) as well as currents (Steer et al., 2020; Pizzo et al., 2023), and quantifying associated rogue wave statistics are promising experimental research venues to be further explored.

Moreover, machine learning will have a higher impact in this field of research (Chapter 8), particularly, when the analytical governing equations reach their limit in modeling and predicting complex wave–structure interactions, wave transformation, and severe wave-breaking processes (Gramstad and Bitner-Gregersen, 2019; Guth and Sapsis, 2019; Häfner et al., 2021; Eeltink et al., 2022).

The rapid development of spatiotemporal imaging techniques (Cobelli et al., 2009; Bergamasco et al., 2017) will improve wave data acquisition, evaluation, and analysis of directional wave hydrodynamics. These will also play a key role in the ascertainment of the dominant wave-focusing mechanism responsible for the spontaneous rogue wave emergence for underlying sea states, in the laboratory and the ocean.

References

Adcock, T.A., Taylor, P.H., 2014. The physics of anomalous 'rogue' ocean waves. Reports on Progress in Physics 77, 105901.

Akhmediev, N., Ankiewicz, A., 1997. Nonlinear Pulses and Beams. Springer.

Alberello, A., Chabchoub, A., Monty, J.P., Nelli, F., Lee, J.H., Elsnab, J., et al., 2018. An experimental comparison of velocities underneath focussed breaking waves. Ocean Engineering 155, 201–210.

Babanin, A., 2011. Breaking and Dissipation of Ocean Surface Waves. Cambridge University Press.

Bacot, V., Labousse, M., Eddi, A., Fink, M., Fort, E., 2016. Time reversal and holography with spacetime transformations. Nature Physics 12, 972–977.

Baldock, T., Swan, C., Taylor, P., 1996. A laboratory study of nonlinear surface waves on water. Philosophical Transactions of the Royal Society of London. Series A: Mathematical, Physical and Engineering Sciences 354, 649–676.

Benjamin, T.B., 1967. Instability of periodic wave trains in nonlinear dispersive systems. Proceedings of the Royal Society of London. Series A. Mathematical and Physical Sciences 299, 59–76.

Benjamin, T.B., Feir, J.E., 1967. The disintegration of wave trains on deep water part 1. theory. Journal of Fluid Mechanics 27, 417–430.

Bergamasco, F., Torsello, A., Sclavo, M., Barbariol, F., Benetazzo, A., 2017. Wass: an open-source pipeline for 3D stereo reconstruction of ocean waves. Computers & Geosciences 107, 28–36.

Cavaleri, L., Barbariol, F., Benetazzo, A., Bertotti, L., Bidlot, J.R., Janssen, P., et al., 2016. The draupner wave: a fresh look and the emerging view. Journal of Geophysical Research: Oceans 121, 6061–6075.

Chabchoub, A., Fink, M., 2014. Time-reversal generation of rogue waves. Physical Review Letters 112, 124101.

Chabchoub, A., Hoffmann, N., Akhmediev, N., 2011. Rogue wave observation in a water wave tank. Physical Review Letters 106, 204502.

Chabchoub, A., Hoffmann, N., Branger, H., Kharif, C., Akhmediev, N., 2013. Experiments on wind-perturbed rogue wave hydrodynamics using the peregrine breather model. Physics of Fluids 25, 101704.

Chabchoub, A., Kibler, B., Dudley, J.M., Akhmediev, N., 2014. Hydrodynamics of periodic breathers. Philosophical Transactions of the Royal Society A: Mathematical, Physical and Engineering Sciences 372, 20140005.

Chabchoub, A., Mozumi, K., Hoffmann, N., Babanin, A.V., Toffoli, A., Steer, J.N., et al., 2019. Directional soliton and breather beams. Proceedings of the National Academy of Sciences 116, 9759–9763.

Chabchoub, A., Waseda, T., Kibler, B., Akhmediev, N., 2017. Experiments on higher-order and degenerate Akhmediev breather-type rogue water waves. Journal of Ocean Engineering and Marine Energy 3, 385–394.

Clauss, G.F., Klein, M., 2011. The new year wave in a seakeeping basin: generation, propagation, kinematics and dynamics. Ocean Engineering 38, 1624–1639.

Clauss, G.F., Klein, M., Onorato, M., 2011. Formation of extraordinarily high waves in space and time. In: International Conference on Offshore Mechanics and Arctic Engineering, pp. 417–429.

Cobelli, P.J., Maurel, A., Pagneux, V., Petitjeans, P., 2009. Global measurement of water waves by fourier transform profilometry. Experiments in fluids 46, 1037–1047.

de Mello, P., Pérez, N., Adamowski, J., Nishimoto, K., 2016. Wave focalization in a wave tank by using time reversal technique. Ocean Engineering 123, 314–326.

Dematteis, G., Grafke, T., Onorato, M., Vanden-Eijnden, E., 2019. Experimental evidence of hydrodynamic instantons: the universal route to rogue waves. Physical Review X 9, 041057.

Deng, Y., Yang, J., Tian, X., Li, X., 2016. Experimental investigation on rogue waves and their impacts on a vertical cylinder using the peregrine breather model. Ships and Offshore Structures 11, 757–765.

Draycott, S., Stansby, P., McAllister, M., Davey, T., Jordan, L., Tosdevin, T., et al., 2022. The numerical re-creation of experimentally generated nonlinear irregular wave fields using a time-reversal approach. Applied Ocean Research 129, 103397.

Ducrozet, G., Bonnefoy, F., Mori, N., Fink, M., Chabchoub, A., 2020. Experimental reconstruction of extreme sea waves by time reversal principle. Journal of Fluid Mechanics 884, A20.

Ducrozet, G., Fink, M., Chabchoub, A., 2016. Time-reversal of nonlinear waves: applicability and limitations. Physical Review Fluids 1, 054302.

Dudley, J.M., Genty, G., Mussot, A., Chabchoub, A., Dias, F., 2019. Rogue waves and analogies in optics and oceanography. Nature Reviews Physics 1, 675–689.

Dysthe, K., Krogstad, H.E., Müller, P., 2008. Oceanic rogue waves. Annual Review of Fluid Mechanics 40, 287–310.

Dysthe, K.B., 1979. Note on a modification to the nonlinear Schrödinger equation for application to deep water waves. Proceedings of the Royal Society of London. A. Mathematical and Physical Sciences 369, 105–114.

Dysthe, K.B., Trulsen, K., 1999. Note on breather type solutions of the NLS as models for freak-waves. Physica Scripta T82, 48–52.

Eeltink, D., Armaroli, A., Luneau, C., Branger, H., Brunetti, M., Kasparian, J., 2020. Separatrix crossing and symmetry breaking in NLSE-like systems due to forcing and damping. Nonlinear Dynamics 102, 2385–2398.

Eeltink, D., Branger, H., Luneau, C., He, Y., Chabchoub, A., Kasparian, J., et al., 2022. Nonlinear wave evolution with data-driven breaking. Nature Communications 13, 2343.

Eeltink, D., Lemoine, A., Branger, H., Kimmoun, O., Kharif, C., Carter, J., et al., 2017. Spectral up-and downshifting of Akhmediev breathers under wind forcing. Physics of Fluids 29, 107103.

Fedele, F., Brennan, J., Ponce de Léon, S., Dudley, J., Dias, F., 2016. Real world ocean rogue waves explained without the modulational instability. Scientific rReports 6, 27715.

Fink, M., 1997. Time reversed acoustics. Physics Today 50, 34–40.

Gemmrich, J., Cicon, L., 2022. Generation mechanism and prediction of an observed extreme rogue wave. Scientific Reports 12, 1–10.

Geva, M., Shemer, L., 2022. Excitation of initial waves by wind: a theoretical model and its experimental verification. Physical Review Letters 128, 124501.

Gramstad, O., Bitner-Gregersen, E., 2019. Predicting extreme waves from wave spectral properties using machine learning. International Conference on Offshore Mechanics and Arctic Engineering. American Society of Mechanical Engineers, p. V003T02A005.

Guth, S., Sapsis, T.P., 2019. Machine learning predictors of extreme events occurring in complex dynamical systems. Entropy 21, 925.

Häfner, D., Gemmrich, J., Jochum, M., 2021. Real-world rogue wave probabilities. Scientific Reports 11, 10084.

He, Y., Slunyaev, A., Mori, N., Chabchoub, A., 2022b. Experimental evidence of nonlinear focusing in standing water waves. Physical Review Letters 129, 144502.

He, Y., Witt, A., Trillo, S., Chabchoub, A., Hoffmann, N., 2022a. Extreme wave excitation from localized phase-shift perturbations. Physical Review E 106, L043101.

Holthuijsen, L.H., 2010. Waves in Oceanic and Coastal Waters. Cambridge University Press.

Houtani, H., Waseda, T., Tanizawa, K., 2018. Experimental and numerical investigations of temporally and spatially periodic modulated wave trains. Physics of Fluids 30, 034101.

Iafrati, A., Babanin, A., Onorato, M., 2013. Modulational instability, wave breaking, and formation of large-scale dipoles in the atmosphere. Physical Review Letters 110, 184504.

Karjanto, N., Van Groesen, E., 2010. Qualitative comparisons of experimental results on deterministic freak wave generation based on modulational instability. Journal of Hydro-Environment Research 3, 186–192.

Karmpadakis, I., Swan, C., Christou, M., 2019. Laboratory investigation of crest height statistics in intermediate water depths. Proceedings of the Royal Society 475A, 20190183.

Kharif, C., Giovanangeli, J.P., Touboul, J., Grare, L., Pelinovsky, E., 2008. Influence of wind on extreme wave events: experimental and numerical approaches. Journal of Fluid Mechanics 594, 209–247.

Kharif, C., Pelinovsky, E., 2003. Physical mechanisms of the rogue wave phenomenon. European Journal of Mechanics-B/Fluids 22, 603–634.

Kharif, C., Pelinovsky, E., Slunyaev, A., 2009. Rogue Waves in the Ocean. Springer Science & Business Media.

Kibler, B., Fatome, J., Finot, C., Millot, G., Dias, F., Genty, G., et al., 2010. The peregrine soliton in nonlinear fibre optics. Nature Physics 6, 790–795.

Klein, M., Clauss, G.F., Rajendran, S., Soares, C.G., Onorato, M., 2016. Peregrine breathers as design waves for wave-structure interaction. Ocean Engineering 128, 199–212.

Lake, B.M., Yuen, H.C., Rungaldier, H., Ferguson, W.E., 1977. Nonlinear deep-water waves: theory and experiment. Part 2. Evolution of a continuous wave train. Journal of Fluid Mechanics 83, 49–74.

Latheef, M., Swan, C., 2013. A laboratory study of wave crest statistics and the role of directional spreading, Proceedings of the Royal Society, 469A. p. 20120696.

Liao, B., Ma, Y., Ma, X., Dong, G., 2018. Experimental study on the evolution of peregrine breather with uniform-depth adverse currents. Physical Review E 97, 053102.

Longuet-Higgins, M., 1974. Breaking waves in deep or shallow water. Proceedings of the 10th Conference on Naval Hydrodynamics. MIT, p. 605.

Ma, Y., Tai, B., Dong, G., Fu, R., Perlin, M., 2022a. An experiment on reconstruction and analysis of in-situ measured freak waves. Ocean Engineering 244, 110312.

Ma, Y., Zhang, J., Chen, Q., Tai, B., Dong, G., Xie, B., et al., 2022b. Progresses in the research of oceanic freak waves: mechanism, modeling, and forecasting. International Journal of Ocean and Coastal Engineering 4, 2250002.

Malila, M.P., Barbariol, F., Benetazzo, A., Breivik, Ø., Magnusson, A.K., Thomson, J., et al., 2023. Statistical and dynamical characteristics of extreme wave crests assessed with field measurements from the North Sea. Journal of Physical Oceanography 53, 509–531.

McAllister, M.L., Draycott, S., Adcock, T., Taylor, P., Van Den Bremer, T., 2019. Laboratory recreation of the Draupner wave and the role of breaking in crossing seas. Journal of Fluid Mechanics 860, 767–786.

Michel, G., Bonnefoy, F., Ducrozet, G., Falcon, E., 2022. Statistics of rogue waves in isotropic wave fields. Journal of Fluid Mechanics 943, A26.

Michel, G., Bonnefoy, F., Ducrozet, G., Prabhudesai, G., Cazaubiel, A., Copie, F., et al., 2020. Emergence of peregrine solitons in integrable turbulence of deep water gravity waves. Physical Review Fluids 5, 082801.

Onorato, M., Osborne, A.R., Serio, M., Cavaleri, L., Brandini, C., Stansberg, C.T., 2006. Extreme waves, modulational instability and second order theory: wave flume experiments on irregular waves. European Journal of Mechanics-B/Fluids 25, 586–601.

Onorato, M., Proment, D., Clauss, G., Klein, M., 2013. Rogue waves: from nonlinear Schrödinger breather solutions to sea-keeping test. PLoS One 8, e54629.

Onorato, M., Waseda, T., Toffoli, A., Cavaleri, L., Gramstad, O., Janssen, P., et al., 2009. Statistical properties of directional ocean waves: the role of the modulational instability in the formation of extreme events. Physical Review Letters 102, 114502.

Osborne, A., 2010. Nonlinear Ocean Waves and the Inverse Scattering Transform. Academic Press.

Peregrine, D.H., 1983. Water waves, nonlinear Schrödinger equations and their solutions. The ANZIAM Journal 25, 16–43.

Peric, R., Hoffmann, N., Chabchoub, A., 2015. Initial wave breaking dynamics of peregrine-type rogue waves: a numerical and experimental study. European Journal of Mechanics-B/Fluids 49, 71–76.

Pierson Jr, W.J., 1955. Wind generated gravity waves, Advances in Geophysics, 2. Elsevier, pp. 93–178.

Pizzo, N., Lenain, L., Rømcke, O., Ellingsen, S.A., Smeltzer, B.K., 2023. The role of lagrangian drift in the geometry, kinematics and dynamics of surface waves. Journal of Fluid Mechanics 954, R4.

Przadka, A., Feat, S., Petitjeans, P., Pagneux, V., Maurel, A., Fink, M., 2012. Time reversal of water waves. Physical Review Letters 109, 064501.

Sapsis, T.P., 2021. Statistics of extreme events in fluid flows and waves. Annual Review of Fluid Mechanics 53, 85–111.
Shemer, L., Alperovich, L., 2013. Peregrine breather revisited. Physics of Fluids 25, 051701.
Shemer, L., Liberzon, D., 2014. Lagrangian kinematics of steep waves up to the inception of a spilling breaker. Physics of Fluids 26, 016601.
Shrira, V.I., Geogjaev, V.V., 2010. What makes the Peregrine soliton so special as a prototype of freak waves? Journal of Engineering Mathematics 67, 11–22.
Slunyaev, A., Pelinovsky, E., Sergeeva, A., Chabchoub, A., Hoffmann, N., Onorato, M., et al., 2013. Super-rogue waves in simulations based on weakly nonlinear and fully nonlinear hydrodynamic equations. Physical Review E 88, 012909.
Steer, J.N., Borthwick, A.G., Onorato, M., Chabchoub, A., Van Den Bremer, T.S., 2019. Hydrodynamic x waves. Physical Review Letters 123, 184501.
Steer, J.N., Borthwick, A.G., Stagonas, D., Buldakov, E., van den Bremer, T.S., 2020. Experimental study of dispersion and modulational instability of surface gravity waves on constant vorticity currents. Journal of Fluid Mechanics 884, A40.
Suret, P., Tikan, A., Bonnefoy, F., Copie, F., Ducrozet, G., Gelash, A., et al., 2020. Nonlinear spectral synthesis of soliton gas in deep-water surface gravity waves. Physical Review Letters 125, 264101.
Toffoli, A., Proment, D., Salman, H., Monbaliu, J., Frascoli, F., Dafilis, M., et al., 2017. Wind generated rogue waves in an annular wave flume. Physical Review Letters 118, 144503.
Tulin, M.P., 1996. Breaking of ocean waves and downshifting. Waves and Nonlinear Processes in Hydrodynamics 177–190.
Tulin, M.P., Waseda, T., 1999. Laboratory observations of wave group evolution, including breaking effects. Journal of Fluid Mechanics 378, 197–232.
van den Bremer, T.S., Whittaker, C., Calvert, R., Raby, A., Taylor, P.H., 2019. Experimental study of particle trajectories below deep-water surface gravity wave groups. Journal of Fluid Mechanics 879, 168–186.
Vanderhaegen, G., Naveau, C., Szriftgiser, P., Kudlinski, A., Conforti, M., Mussot, A., et al., 2021. "Extraordinary" modulation instability in optics and hydrodynamics. Proceedings of the National Academy of Sciences 118, e2019348118.
Waseda, T., Kinoshita, T., Tamura, H., 2009. Evolution of a random directional wave and freak wave occurrence. Journal of Physical Oceanography 39, 621–639.
Waseda, T., Watanabe, S., Fujimoto, W., Nose, T., Kodaira, T., Chabchoub, A., 2021. Directional coherent wave group from an assimilated non-linear wavefield. Frontiers in Physics 9, 622303.
Zakharov, V.E., 1968. Stability of periodic waves of finite amplitude on the surface of a deep fluid. Journal of Applied Mechanics and Technical Physics 9, 190–194.

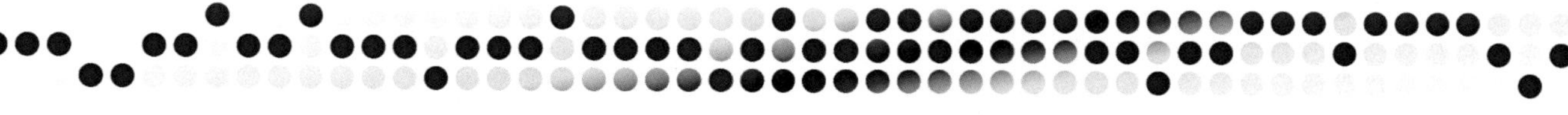

Nomenclature

$H > 2H_{1/3}$	General definition of freak wave
BFI	Benjamin–Feir index
μ_3	Skewness
μ_4	Kurtosis
κ_4	Fourth-order cumulant
$T_{1,2,3,4}$	Fourth-order nonlinear interaction term
σ	Standard deviation
t	Time
$1-D$, $2-D$, $3-D$	Dimensionality of space
(x,y)	Axes of a coordinate horizontal plane
L	Wavelength
$k = (k_x, k_y)$	Wave number
N	Wave action density
f	Frequency
ω	Angular frequency
H	Trough-to-crest wave height
H_{max}	Maximum wave height
R	Two-dimensional surface region
A	Area

One-dimensional temporal interval

η_{max}	Maximum crest height
η	Surface elevation
$S(f, k_x, k_y)$, $S(f)$	Wave spectrum
$H_{1/3}$	Significant wave height based on zero-crossing method
H_s	Significant wave height by spectra or standard deviation
T	Trough-to-crest wave period
$T_{1/3}$	Significant wave period based on zero-crossing method
T_s	Significant wave period by spectra or standard deviation
ϕ	Velocity potential
u, v	Velocity
$m_{i,j,l}$	Moments of the directional spectrum
σ_θ	Directional spreading

Index

Note: Page numbers followed by "*f*" and "*t*" refer to figures and tables, respectively.

Index

O

P

Index

W

X

Z

Printed in the United States
by Baker & Taylor Publisher Services